EXAMKRACKERS

MCAT

Chemistry

3rd Edition

Osote

Publishing

 Published in the United States of America by Osote Publishing, New Jersey.

ISBN 1-893858-16-2 (Volume 4)
ISBN 1-893858-20-0 (5 Volume Set)

3rd Edition

To purchase additional copies of this book or the rest of the 5 volume set, call 1-888-572-2536 or fax orders to 1-201-996-1153.

examkrackers.com

osote.com

audioosmosis.com

Inside cover design consultant: Fenwick Design Inc. (212) 246-9722; (201) 944-4337

Printed in Hong Kong

Acknowledgements

Although I am the author, the hard work and expertise of many individuals contributed to this book. The idea of writing in two voices, a science voice and an MCAT voice, was the creative brainchild of my imaginative friend Jordan Zaretsky. I would like to thank Scott Calvin for lending his exceptional science talent and pedagogic skills to this project. I also must thank three years worth of EXAMKRACKERS students for doggedly questioning every explanation, every sentence, every diagram, and every punctuation mark in the book. Finally, I wish to thank my wife, Silvia, for her support during the difficult times in the past and those that lie ahead.

Read This Section First!

This manual contains all the organic chemistry tested on the MCAT and more. It contains more chemistry than is tested on the MCAT because a deeper understanding of basic scientific principles is often gained through more advanced study. Also, the MCAT often presents passages with imposing topics that may intimidate the test-taker. Some familiarity with these topics will increase the confidence of the test-taker. In order to answer questions quickly and efficiently, it is vital that the test-taker understand what is, and is not, tested directly by the MCAT. To assist the test-taker in gaining this knowledge, this manual will use the following conventions. Any term or concept which is tested directly by the MCAT will be written in **bold and underlined**. To ensure a perfect score on the MCAT, you should thoroughly understand all terms and concepts that are in bold and underlined in this manual. Sometimes it is not necessary to memorize the name of a concept, but it is necessary to understand the concept itself. These concepts will also be in bold and underlined. It is important to note that the converse of the above is not true: just because a topic is not in bold and underlined, does not mean that it is not important.

Any formulae that must be memorized will be written in **Bold and large type**.

If a topic is discussed purely as background knowledge, it will be written in *italics*. If a topic is written in italics, it is not required knowledge for the MCAT but may be explained by an MCAT passage. Answers to questions that directly test knowledge of these topics will always be found in an MCAT passage.

Salty

Text written in this font is me, Salty the Kracker. I will remind you what is and is not an absolute must for MCAT. I will help you develop your MCAT intuition. In addition, I will offer mnemonics, simple methods of viewing a complex concept, and some badly needed comic relief. Don't ignore me, even if you think I am not funny, because my comedy is designed to help you understand and remember. If you think I am funny, tell the boss. I could use a raise.

Each chapter in this manual should be read three times: twice before the class lecture, and once immediately following the lecture. During the first reading, you should not write in the book. Instead, read purely for enjoyment. During the second reading, you should both highlight and take notes in the margins. The last reading should be slow and thorough.

The 15 questions in each lecture should be worked during the second reading before coming to class.

The in-class exams in the back of the book are to be done in class after the lecture. Do not look at them before class. For a rough idea of your MCAT performance on each topic, you can find your scaled score from the chart provided on the answer page.

Warning: Just attending the class will not raise your score. **You must do the work**.

Table of Contents

Class Lectures

Lecture 1: Atoms, Molecules, and Quantum Mechanics........1
Lecture 2: Gases, Kinetics, and Chemical Equilibrium........23
Lecture 3: Thermodynamics........49
Lecture 4: Solutions........73
Lecture 5: Heat Capacity, Phase Change, and Colligative Properties........91
Lecture 6: Acids and Bases........107
Lecture 7: Electrochemistry........127

In-class Exams

Exam 1:........145
Exam 2:........151
Exam 3:........157
Exam 4:........163
Exam 5:........169
Exam 6:........175
Exam 7:........181

Answers to In-class Exams

Answers........186

Explanations to In-class Exams

Exam 1........187
Exam 2........188
Exam 3........190
Exam 4........193
Exam 5........195
Exam 6........196
Exam 7........197

PHYSICAL SCIENCES

DIRECTIONS: Most questions in the Physical Sciences test are organized into groups, each preceded by a descriptive passage. After studying the passage, select the one best answer to each question in the group. Some questions are not based on a descriptive passage and are also independent of each other. You must also select the one best answer to these questions. If you are not certain of an answer, eliminate the alternatives that you know to be incorrect and then select an answer from the remaining alternatives. Indicate your selection by blackening the corresponding oval on your answer document. A periodic table is provided for your use. You may consult it whenever you wish.

PERIODIC TABLE OF THE ELEMENTS

1 H 1.0																	2 He 4.0
3 Li 6.9	4 Be 9.0											5 B 10.8	6 C 12.0	7 N 14.0	8 O 16.0	9 F 19.0	10 Ne 20.2
11 Na 23.0	12 Mg 24.3											13 Al 27.0	14 Si 28.1	15 P 31.0	16 S 32.1	17 Cl 35.5	18 Ar 39.9
19 K 39.1	20 Ca 40.1	21 Sc 45.0	22 Ti 47.9	23 V 50.9	24 Cr 52.0	25 Mn 54.9	26 Fe 55.8	27 Co 58.9	28 Ni 58.7	29 Cu 63.5	30 Zn 65.4	31 Ga 69.7	32 Ge 72.6	33 As 74.9	34 Se 79.0	35 Br 79.9	36 Kr 83.8
37 Rb 85.5	38 Sr 87.6	39 Y 88.9	40 Zr 91.2	41 Nb 92.9	42 Mo 95.9	43 Tc (98)	44 Ru 101.1	45 Rh 102.9	46 Pd 106.4	47 Ag 107.9	48 Cd 112.4	49 In 114.8	50 Sn 118.7	51 Sb 121.8	52 Te 127.6	53 I 126.9	54 Xe 131.3
55 Cs 132.9	56 Ba 137.3	57 La* 138.9	72 Hf 178.5	73 Ta 180.9	74 W 183.9	75 Re 186.2	76 Os 190.2	77 Ir 192.2	78 Pt 195.1	79 Au 197.0	80 Hg 200.6	81 Tl 204.4	82 Pb 207.2	83 Bi 209.0	84 Po (209)	85 At (210)	86 Rn (222)
87 Fr (223)	88 Ra 226.0	89 Ac† 227.0	104 Unq (261)	105 Unp (262)	106 Unh (263)	107 Uns (262)	108 Uno (265)	109 Une (267)									

*	58 Ce 140.1	59 Pr 140.9	60 Nd 144.2	61 Pm (145)	62 Sm 150.4	63 Eu 152.0	64 Gd 157.3	65 Tb 158.9	66 Dy 162.5	67 Ho 164.9	68 Er 167.3	69 Tm 168.9	70 Yb 173.0	71 Lu 175.0
†	90 Th 232.0	91 Pa (231)	92 U 238.0	93 Np (237)	94 Pu (244)	95 Am (243)	96 Cm (247)	97 Bk (247)	98 Cf (251)	99 Es (252)	100 Fm (257)	101 Md (258)	102 No (259)	103 Lr (260)

Lecture 1: Atoms, Molecules, and Quantum Mechanics

Atoms

All mass consists of tiny particles called **atoms**. Each atom is composed of a **nucleus** surrounded by one or more electrons. The nucleus contains **protons** and **neutrons**, collectively called *nucleons*, held together by the *strong nuclear force*. (More precisely, the strong nuclear force holds together the three *quarks* that make up each nucleon, and it is the 'spill over' from this force that holds together the nucleons.) Protons and neutrons are approximately equal in size and mass. Protons have a positive charge and neutrons are electrically neutral.

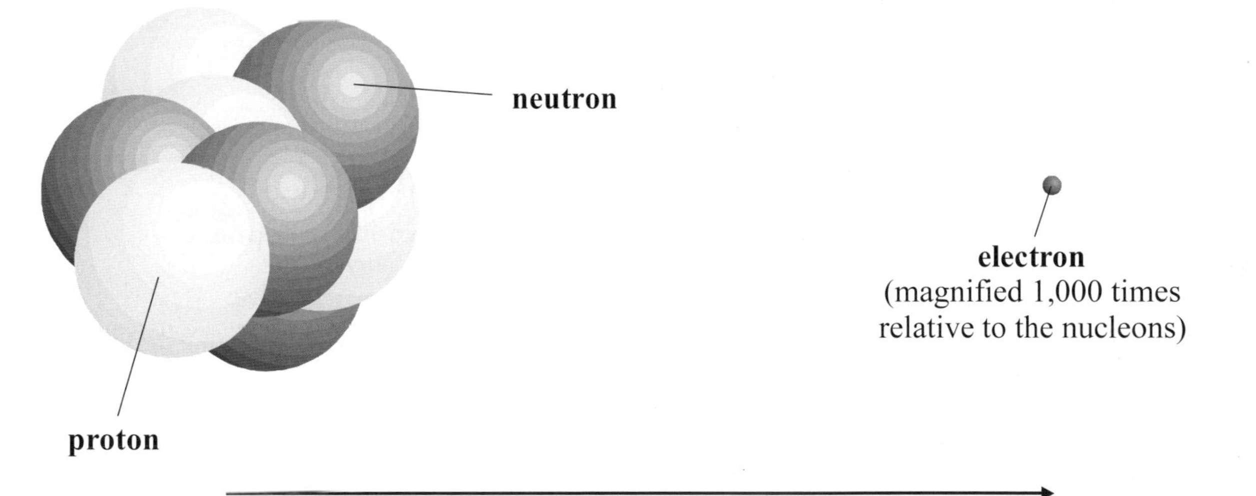

Surrounding the nucleus, at great distances compared to a nucleon radius, are **electrons**. Electrons are very small compared to nucleons. Electrons and protons have opposite charges of equal magnitude. Although for convenience we often think of the charge on an electron as –1 and the charge on a proton as +1, we should remember that this charge is in electron units of '***e***'. A charge of 1 *e* is equal to 1.6×10^{-19} coulombs. If an atom is electrically neutral, it contains the same number of protons as electrons. If it is not electrically neutral, it is called an **ion**. Positively charged ions are called **cations**, and negatively charged ions are called **anions**.

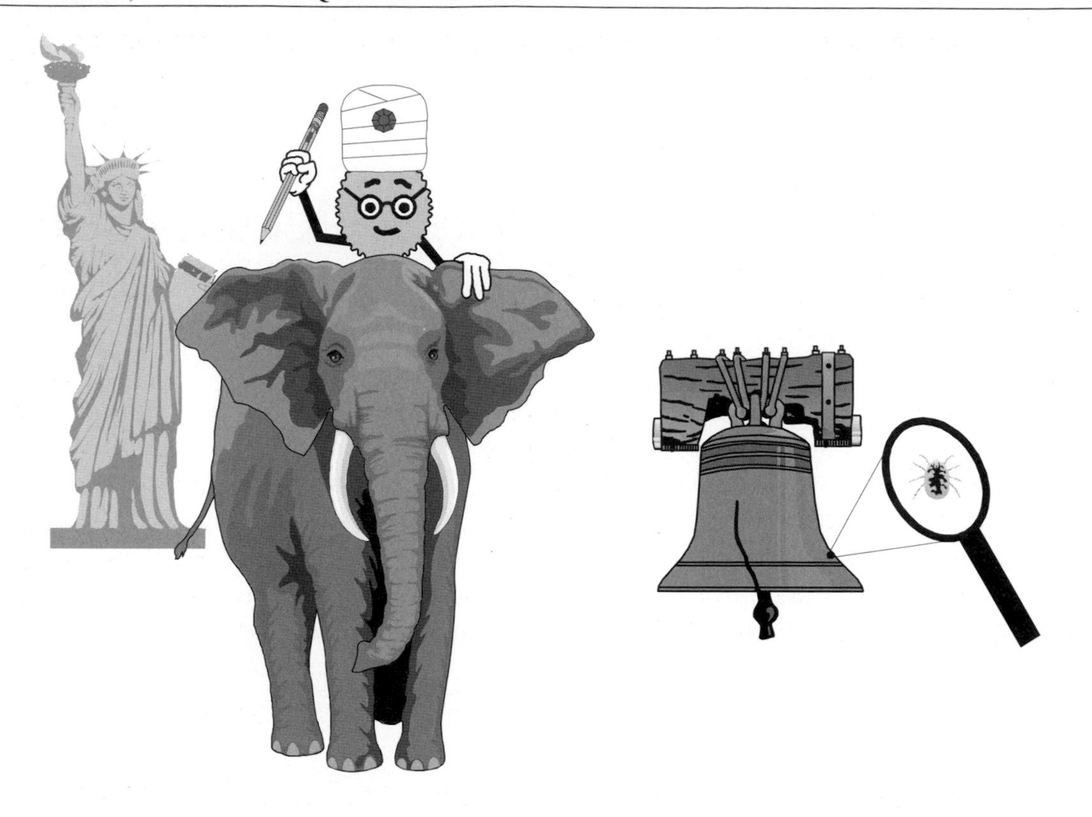

If a proton
were the size
of an elephant in New York

the electron would be
the size
of a tick in Philadelphia

Any single atom must be one of the many **elements**. Elements are the building blocks of all compounds and cannot be decomposed into simpler substances by chemical or physical means. Any element can be displayed as follows:

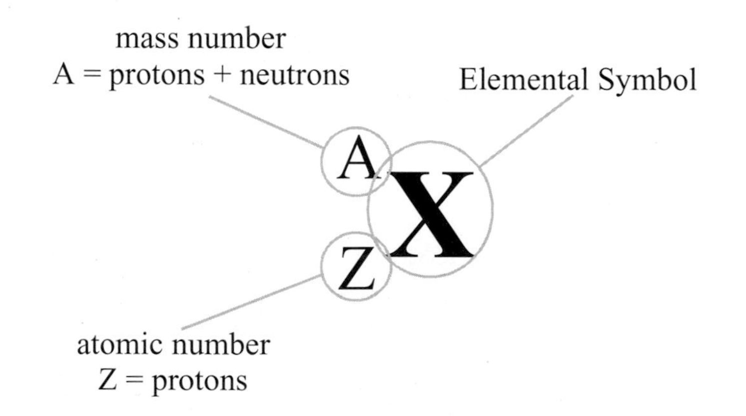

where **A** is the **mass number** or number of protons plus neutrons, and **Z** is the **atomic number** or number of protons. The atomic number is the identity number of any element. If we know the atomic number, then we know the element. This is not true of the mass number or the number of electrons. Any element may have any number of neutrons or electrons, but only one number of protons.

Two or more atoms of the same element that contain different numbers of neutrons are said to be **isotopes**. Examples of three isotopes for carbon are:

$$^{12}\mathbf{C},\ ^{13}\mathbf{C} \text{ and } ^{14}\mathbf{C}.$$

Each of these isotopes contains 6 protons. 6 protons define carbon. ^{12}C contains 6 neutrons, ^{13}C contains 7 neutrons, and ^{14}C contains 8 neutrons. The number of electrons contained by these isotopes will depend upon whether or not a given atom of the isotope has a charge. The number of electrons will not change the identity of the element.

Although the mass number is a good approximation of the mass of an atom, it is not exact. The **atomic weight** of an atom is given in **atomic mass units (amu)**. (The atomic weight of an element when given in 'amu's is actually a mass and not a weight.) An amu is defined by ^{12}C. By definition, one atom of ^{12}C has an atomic weight of 12 amu. All other atomic weights are measured against this standard. Since carbon naturally occurs as a mixture of its isotopes, the atomic weight of carbon is listed as the weighted average of its isotopes or 12.011 amu. (This is very close to 12 amu because almost 99% of carbon occurs in nature as ^{12}C.)

Think of an amu as approximately the mass of one proton or one neutron.

^{12}C also defines a **mole**. A mole (or **Avogadro's number, 6.022×10^{23}**) is the number of carbon atoms in 12 grams of ^{12}C. This means that we can read atomic weights from the periodic table as either amu or g/mol. Thus if we are given the amount of an element or compound in grams, we can divide by the atomic or molecular weight to find the number of moles in that sample.

$$\textbf{moles} = \frac{\textbf{grams}}{\textbf{atomic or molecular weight}}$$

The Periodic Table

The periodic table lists the elements from left to right in the order of their atomic numbers. Each horizontal row is called a **period**. The vertical columns are called **groups or families**. There are two methods used to number the groups. The newer method is to number them 1 through 18 from left to right. The older method, which is sometimes still used on the MCAT, is to separate the groups into sections A and B. These sections are then numbered with roman numerals as shown below.

You should recognize the names of the following five groups: alkali metals; alkaline earth metals; halogens; noble gases; and the transition metals. (The transition metals are actually 10 groups.) The section A groups are known as the *representative elements*.

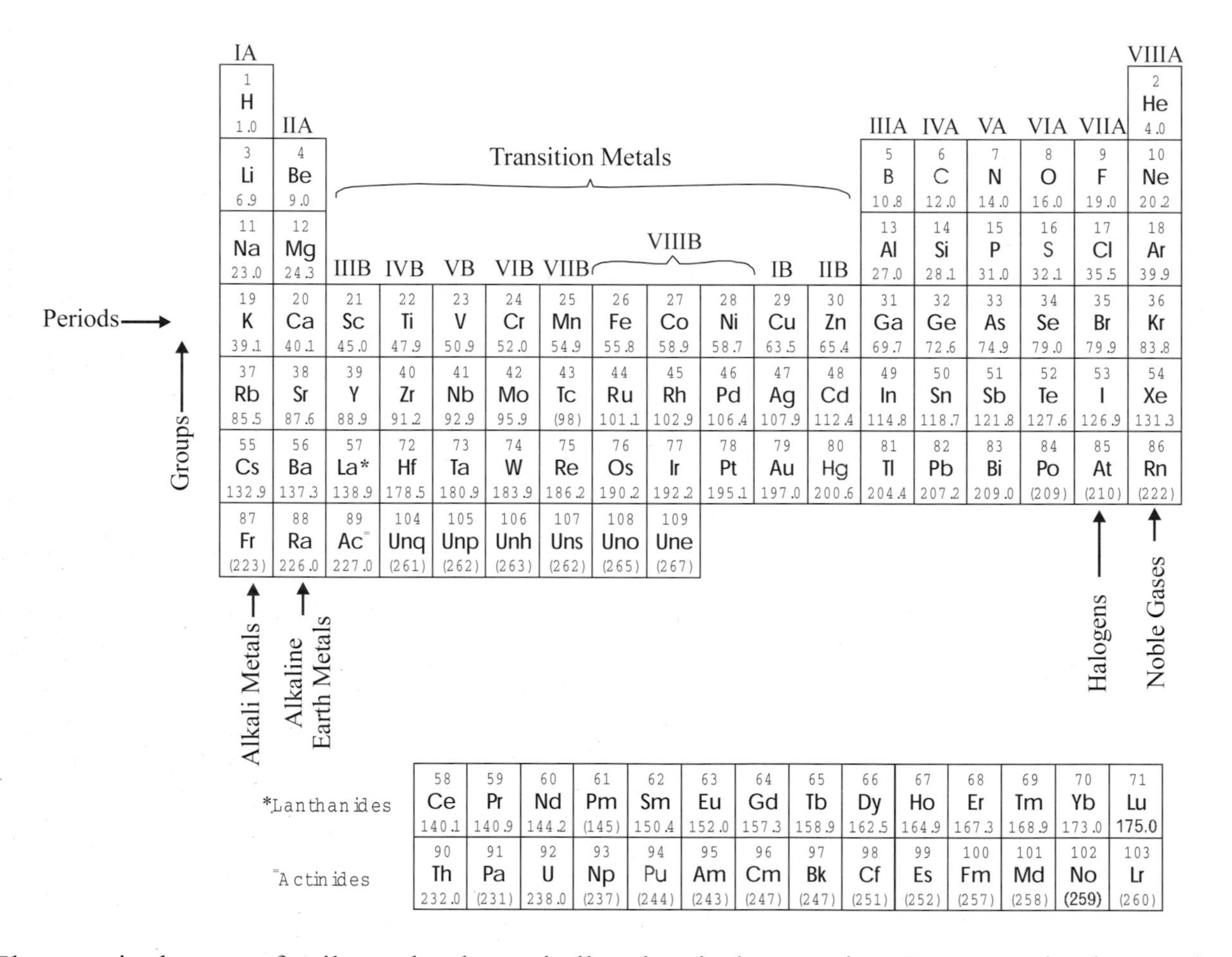

Elements in the same family tend to have similar chemical properties. For example, they tend to make the same number of bonds, and exist as similarly charged ions. You should know that alkali metals tend to exist as +1 ions, alkaline earth metals as +2 ions, and halogens as –1 ions. Noble gases are non-reactive. They are sometimes called the **inert gases**. Metals tend to lose electrons to form positive ions. Metallic character means ductile (easily stretched), malleable (easily hammers into thin strips), and conductive to heat and electricity.

If we assume that we are dealing with neutral atoms, as we move from left to right, one element to the next on the periodic table, we add a proton and an electron with each step. Since there is a limited amount of space near the nucleus of an atom, the electrons are added in layers, one layer over the next, starting near the nucleus and moving outward. These layers are called **shells**. Each new period on the periodic table represents the introduction of a shell. The positively charge nucleus pulls the negatively charged electrons inward. However, the electrons on the outer layers are partially **shielded** from the electric field of the nucleus by the inner shells. The amount of charge that a specific electron *feels* from the nucleus when shielding is considered is called the **effective nuclear charge (Z_{eff})**. As we move from left to right across the periodic table, each additional electron is added to the same or an inward shell (The d orbitals represent a more inward shell.), and, with each electron, a proton is also added. Thus the effective nuclear charge for each new electron generally increases when moving from left to right across the periodic table. Due to shielding, there is a slight decreasing trend in Z_{eff} when a second electron is added to an orbital. Z_{eff} also generally increases going down the periodic table.

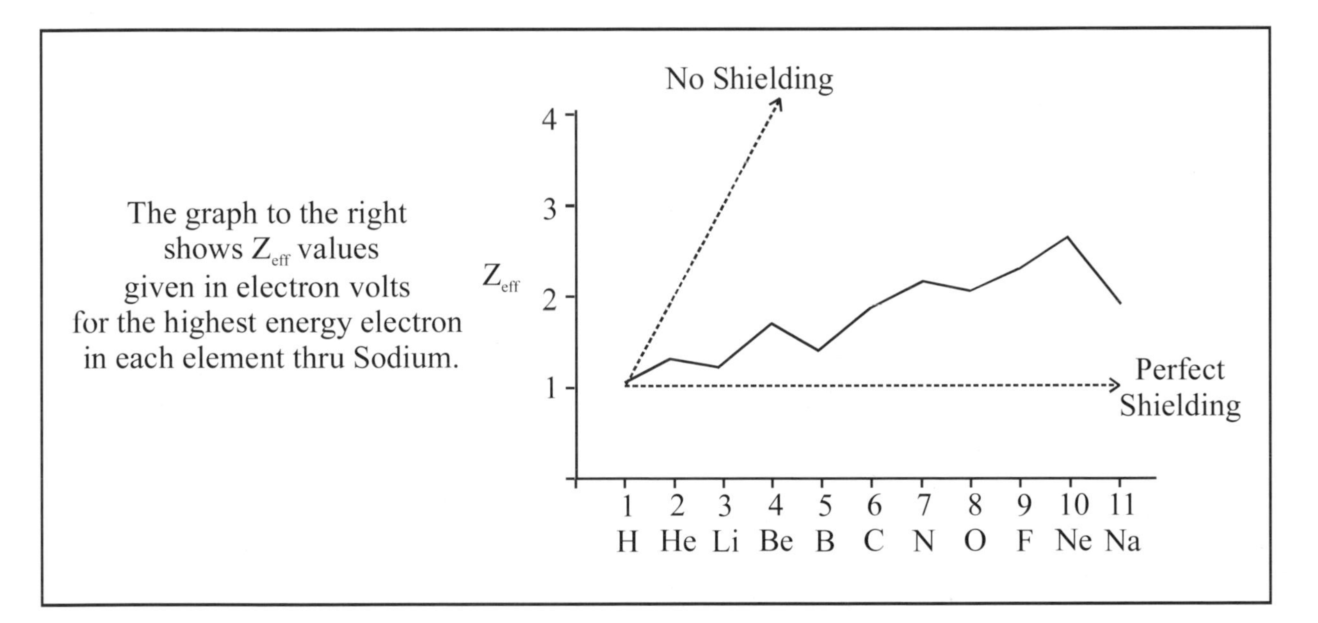

With Z_{eff} in mind, we can make general predictions about the elements based upon their position in the periodic table. These predictions are called **periodic trends**. Since the effective nuclear charge increases when moving from left to right, each additional electron is pulled more strongly toward the nucleus. This results in a smaller **atomic radius**. Of course, with each added shell the atom grows larger. Thus, atomic radius also increases from the top of the periodic table to the bottom.

When an electron is more strongly attached to the nucleus, more energy is required to detach it. The energy necessary to detach an electron from a nucleus is called **ionization energy**. The energy necessary to detach an electron from a neutral atom is called the first ionization energy. (By definition, the atom being ionized is gaseous.) The energy for the removal of a second electron from the same atom is called the **second ionization energy**, and so on. The second ionization energy is always much greater than the first because when one electron is removed, the effective nuclear charge on the other electrons increases. Ionization energy generally increases along the periodic table from left to right and from bottom to top. This trend is explained by Z_{eff}. Z_{eff} increases when moving across a period to the right making it tougher to knock off an electron. Although Z_{eff} also increases when moving down the periodic table, the distance of the electron from the nucleus increases as well, thus decreasing the electric field at the point of the electron. The decreased electric field has less strength to hold the electron to the atom.

Electronegativity is the tendency of an atom to attract an electron in a bond that it shares with another atom. The most commonly used measurement of electronegativity is the *Pauling scale*, which ranges from cesium with a value of 0.79 to fluorine with a value of 4.0. Electronegativity also tends to increase from left to right and bottom to top on the periodic table, and is related to Z_{eff}.

Electron affinity is the willingness of an atom to accept an additional electron. (More precisely, it is the energy released when an electron is added to a gaseous atom. **Warning:** Some books define electron affinity as the energy *absorbed* when an electron is added to a gaseous atom. The difference between the two definitions is one of perspective, and is reflected with a positive or negative sign.) Electron affinity also tends to increase on the periodic table from left to right and from bottom to top, and is related to Z_{eff}. In other words, energy tends to be released when an electron is added to elements positioned to the right and up on the periodic table, and energy tends to be absorbed when an electron is added to elements positioned on the periodic table down and to the left.

The final important periodic trend, **metallic character,** tends to increase from right to left and top to bottom. All these periodic trends are related to the effective nuclear charge.

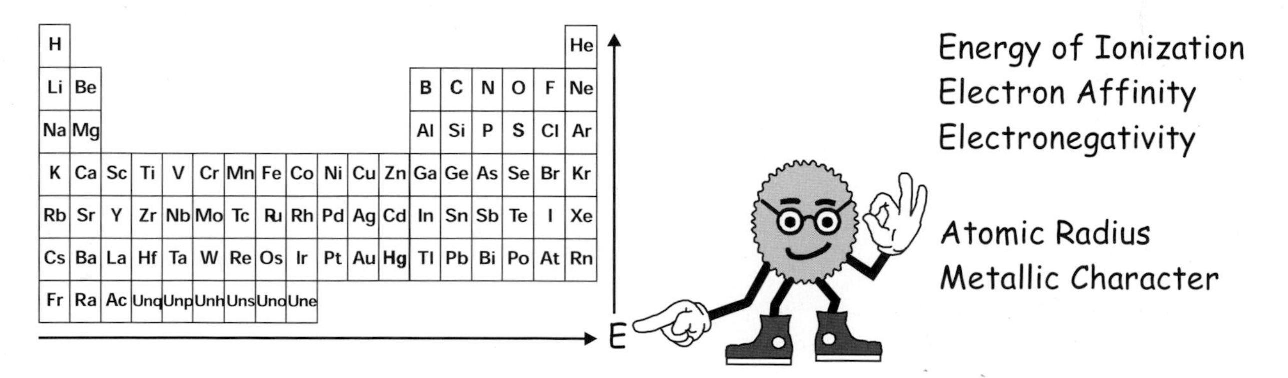

An easy way to remember the 5 periodic trends is as follows: if it begins with an 'E', as shown above, then it increases going to the right and up; if it doesn't begin with an 'E', then it increases in the opposite direction. Be careful! This mnemonic requires that you think of 'ionization energy' as 'energy of ionization' so that it begins with an 'E'. Z_{eff} is not considered a periodic trend for this mnemonic.

Questions 1 through 5 are **NOT** based on a descriptive passage.

1. Which of the following increases with increasing atomic number within a family on the periodic table?

A. electronegativity
B. electron affinity
C. atomic radius
D. ionization energy

2. Which of the following molecules has the greatest dipole moment?

A. H_2
B. O_2
C. HF
D. HBr

3. How many carbon atoms exist in 12 amu of ^{12}C?

A. 1
B. 12
C. 6.02×10^{23}
D. 7.22×10^{24}

4. Which of the following most likely represents the correct order of ion size from greatest to smallest?

A. O^{2-}, F^-, Na^+, Mg^{2+}
B. Mg^{2+}, Na^+, F^-, O^{2-}
C. Na^+, Mg^{2+}, O^{2-}, F^-
D. Mg^{2+}, Na^+, O^{2-}, F^-

5. A natural sample of carbon contains 99% of ^{12}C. How many moles of ^{12}C are likely to be found in a 48.5 gram sample of carbon obtained from nature?

A. 1
B. 4
C. 12
D. 49.5

Answers to Questions 1-5

1. **C is correct.** A family or group is the name for any vertical column on the periodic table. Of the choices given, only atomic radius increases going down a column. Although electron affinity is a possible choice depending upon the definition used, atomic radius is an unambiguous choice.

2. **C is correct.** The dipole moment will be greatest for the atoms with greatest difference in electronegativity. Based upon periodic trends, H and F will have the greatest dipole moment.

3. **A is correct.** By definition there are 12 amu in one atom of ^{12}C.

4. **A is correct.** The number of electrons on each ion is the same, but the nuclear charge increases with increasing atomic number and draws the electrons inward with greater force. This can be used as a general prediction of ion size.

5. **B is correct.** Don't do any complicated calculations. This is the type of problem that everyone will get right, but many will spend too much time trying to be exact. First assume that 100% of the sample is ^{12}C. Now use the formula moles = grams/molecular weight. This is very close to 4. The 1% that is not ^{12}C is insignificant.

Molecules

Atoms can be held together by **bonds**. In one type of bond, two electrons are shared by two nuclei. This is called a covalent bond. The negatively charged electrons are pulled toward both positively charged nuclei by electrostatic forces. This 'tug of war' between the nuclei for the electrons holds the atoms together. If the nuclei come too close to each other, the positively charged nuclei repel each other. These repulsive and attractive forces achieve a balance to create a bond. The diagram to the right compares the internuclear distance between two hydrogen atoms to their electrostatic potential energy level as a system. The **bond length** is defined as the point where the energy level is the lowest. Two atoms will only form a bond if they can lower their overall energy level by doing so. Nature tends to seek the lowest energy state.

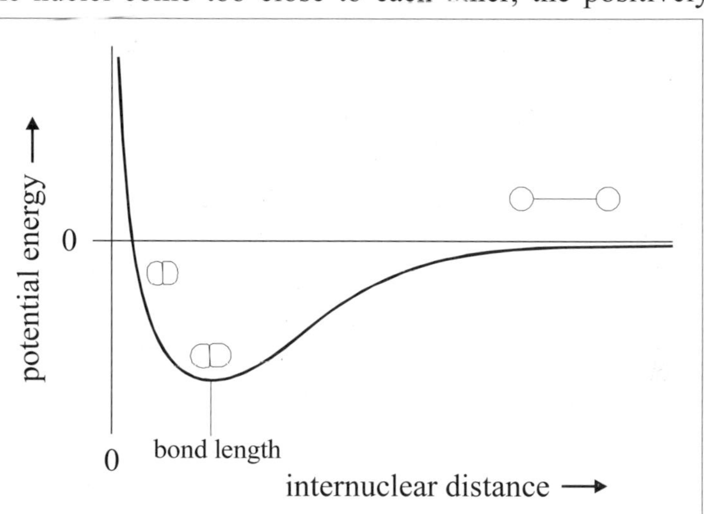

If we separate the atoms by an infinite distance, the forces between them, and thus the energy, go to zero. The energy necessary to achieve a complete separation is given by the vertical distance on the graph between the bond length and zero. This is called **the bond dissociation energy or bond energy.**

Notice from the graph that energy is always required to break a bond. Conversely, no energy is ever gained, liberated, or achieved by breaking a bond. (Energy from ATP is released when the new bonds of ADP and iP are formed, and not when the ATP bonds are broken.)

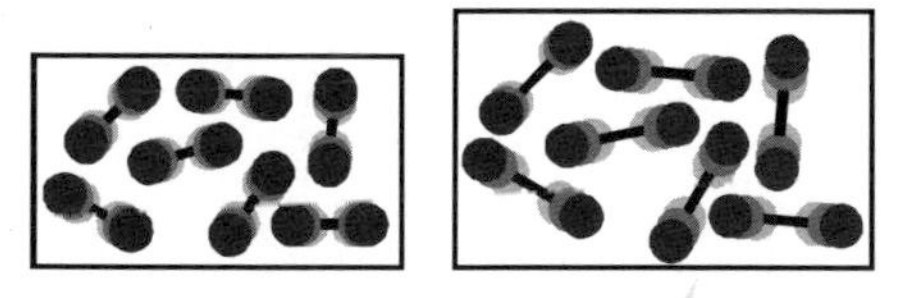

Heated molecules in a solid vibrate more and have a greater average bondlength.

The graph above also explains why most solids expand when they are heated. The bond energy and internuclear distance between the atoms in solids follows this graph. Molecules in a solid vibrate causing bond-length fluctuations (stretching and squeezing of the bond). Heat increases these vibrations. Notice that the slope to the left of the bond length is steeper than to the right. This means that for equal energy increases or vertical movements, there is a greater rightward shift than leftward shift. Thus, there is an increase in the average internuclear distance, and the solid expands.

A collection of atoms of different elements held together by bonds is called a **compound**. In some compounds, groups of atoms form repeated, distinct units called **molecules**. In all pure compounds, the relative number of atoms of one element to another can be represented by a ratio of whole numbers. This ratio is called the **empirical formula**. In molecular compounds, the exact number of elemental atoms in each molecule can be represented by a **molecular formula**. The empirical formula for glucose is CH_2O. The molecular formula is $C_6H_{12}O_6$.

From the empirical formula and atomic weight, we can find the percent composition of a compound by mass. To do this, multiply an atom's atomic weight by the number of atoms it contributes to the empirical formula. Divide your result by the weight of all the atoms in the empirical formula. This gives you the mass fraction of the compound represented by the atom. Now multiply this fraction by 100, and you have the percent composition by mass.

The fraction of carbon present by mass in glucose (empirical formula = CH_2O) is found as follows:

$$\frac{\text{molecular weight of carbon}}{\text{molecular weight of } CH_2O} = \frac{12}{30} = 0.4$$

$$0.4 \times 100 = 40$$

glucose is 40% carbon by mass

To find the empirical formula from the percent mass composition, you assume that you have a 100 gram sample. Now the percent translates directly to grams. When you divide the grams by atomic weight, you get moles. Now divide by the lowest common denominator. This is the number of atoms represented by each element in the empirical formula. In order to find the molecular formula, we would need more information.

If we are asked to find the empirical formula of a compound that is 6% hydrogen and 94% oxygen by mass, we do the following:

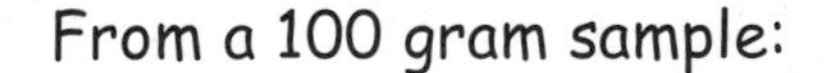

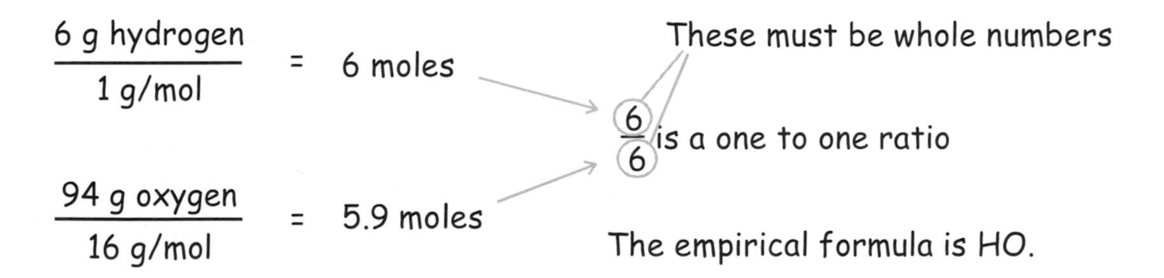

Chemical Reactions and Equations

When a compound undergoes a reaction and maintains its molecular structure and thus its identity, the reaction is called a **physical reaction**. Melting, evaporation, separation of enantiomers, and rotation of polarized light are some examples of physical reactions. When a compound undergoes a reaction and changes its molecular structure to form a new compound, the reaction is called a **chemical reaction**. Combustion, hydrolysis, and substitution are all examples of chemical reactions. Chemical reactions can be represented by a chemical equation with the molecular formulae of the reactants on the left and the products on the right.

$$CH_4 + 2O_2 \rightarrow CO_2 + 2H_2O$$

Notice that there is a conservation of atoms from the left to the right side of the equation. In other words, there is the same number of oxygen, hydrogen, and carbon atoms on the right as on the left. This means the equation is balanced. On the MCAT, if the answer is given in equation form, the correct answer will be a balanced equation unless specifically indicated to the contrary.

In the equation above, O_2 is preceded by a coefficient of 2. A coefficient of 1 is assumed for all molecules not preceded by a coefficient. These coefficients indicate the relative number of molecules. They represent the number of single molecules, moles of molecules, dozens of molecules or any other quantity. They do not represent the mass, the number of grams, or kilograms.

Generally, to say that a reaction **runs to completion** means that it continues to move to the right until the supply of at least one of the reactants is depleted. (Reactions often don't run to completion because they reach equilibrium first. See Chemistry Lecture 2.) As indicated by the equation above, if we were to react 4 moles of methane (CH_4) with 6 moles of oxygen gas (O_2), and the reaction ran to completion, we would be left with 1 mole of methane. Since we would run out of oxygen first, oxygen is our **limiting reagent**. Notice that the limiting reagent is not necessarily the reactant of which there is the least; it is the reactant that would be completely used up if the reaction were to run to completion.

The amount of product produced when a reaction runs to completion is called the **theoretical yield**. As mentioned above, reactions often don't run to completion. The amount of actual product divided by the theoretical yield, times 100, gives the **percent yield**.

$$\frac{\text{actual yield}}{\text{theoretical yield}} \times 100 = \text{percent yield}$$

Bonding in Solids

Solids can be *crystalline or amorphous*. A crystal has a sharp melting point and a characteristic shape with a well ordered structure of repeating units which can be atoms, molecules or ions. A crystal is classified as *ionic, network covalent, metallic*, or *molecular* depending upon the nature of the chemical bonding and the intermolecular forces in the crystal. Ionic crystals consist of oppositely charged ions held together by electrostatic forces. Salts are ionic crystals. Metallic crystals are single metal atoms bonded together by delocalized electrons. These delocalized electrons allow metallic crystals to efficiently conduct heat and electricity. They also make metallic crystals malleable and ductile. Network covalent crystals consist of an infinite network of atoms held together by polar and nonpolar bonds. Diamond and crystal SiO_2 are common examples of network covalent crystals. It is not possible to identify individual molecules in ionic, metallic, and network covalent crystals. Molecular crystals are composed of individual molecules held together by intermolecular bonds. Ice is an example of a molecular crystal.

An amorphous solid has no characteristic shape and melts over a temperature range. Glass (SiO_2) is an amorphous solid. Some substances are capable of forming both crystalline and amorphous solids.

Polymers are solids with repeated structural units. They can be crystalline or amorphous. Generally, rapid cooling of liquid polymers results in amorphous solids and slow cooling results in crystalline solids. Examples of biopolymers are DNA, glycogen, and protein.

The MCAT does not directly test your knowledge of the structure of solids beyond ionic and molecular solids; however, it is good to at least be aware that atoms can form substances in many ways. A recent MCAT had a passage on this topic.

Questions 6 through 10 are **NOT** based on a descriptive passage.

6. What is the empirical formula of a neutral compound containing 58.6% oxygen, 39% sulfur, 2.4% hydrogen by mass?

A. HSO_3^-
B. HSO_4^-
C. H_2SO_3
D. H_2SO_4

7. What is the percent by mass of carbon in CO_2?

A. 12%
B. 27%
C. 33%
D. 44%

8. Sulfur dioxide oxidizes in the presence of O_2 gas as per the reaction:

$$2SO_2(g) + O_2(g) \rightarrow 2SO_3(g)$$

Approximately how many grams of sulfur trioxide are produced by the complete oxidation of 1 mole of sulfur dioxide?

A. 1 g
B. 2 g
C. 80 g
D. 160 g

9. The forces that allow H_2 to form a liquid at low temperatures are best described as:

A. hydrogen bonding.
B. London dispersion forces.
C. covalent bonding.
D. intramolecular forces.

10. When gaseous ammonia is passed over solid copper(II)oxide at high temperatures, nitrogen gas is formed.

$$2NH_3(g) + 3CuO(s) \rightarrow N_2(g) + 3Cu(s) + 3H_2O(g)$$

What is the limiting reagent when 34 grams of ammonia form 26 grams of nitrogen in a reaction that runs to completion?

A. NH_3
B. CuO
C. N_2
D. Cu

Answers to Questions 6-10

6. **C is correct.** We start by assuming a 100 gram sample. By dividing grams by molecular weight, we obtain moles. $58.6/16 \approx 3.6$, $2.4/1 = 2.4$, $39/32 = 1.2$. Now we divide through by the lowest number of moles: $3.6/1.2 = 3$; $2.4/1.2 = 2$; $1.2/1.2 = 1$. This gives you the molar ratio of each element. Just to reduce the necessary calculations, the question tells you that it is a neutral compound. Nevertheless, MCAT questions with this much calculation occasionally come up, but they are few and far between. Maybe three on one entire exam.

7. **B is correct.** C has 12 g/mol and O has 16 g/mol. The total weight of CO_2 is 44 g/mol. Carbon's weight divided by the total weight is $12/44 = 0.27$. We multiply by 100 to get 27%.

8. **C is correct.** When one mole of sulfur dioxide is oxidized, one mole of sulfur trioxide is produced. One mole of sulfur trioxide has a mass of 80 g.

9. **B is correct.** London dispersion forces are the intermolecular forces created by instantaneous dipoles. These are the only forces which allow nonpolar molecules to form condensed phases.

10. **B is correct.** Normally, 34 grams of ammonia (2 moles) could make 28 grams of nitrogen (1 mole), but here, only 26 grams were made. In a reaction that runs to completion, this must be due to lack of CuO.

Quantum Mechanics

The MCAT requires a small amount of knowledge concerning quantum mechanics. Everything that you'll need to know is listed below. Quantum mechanics basically says that elementary particles can only change their energy level in specific, discreet units.

Quantum Numbers

A set of four quantum numbers is the address for an electron in a given atom. No two electrons in the same atom can have the same four quantum numbers.

The first quantum number is the **principal quantum number, n**. The principal quantum number designates the **shell** level. The larger the principal quantum number, the greater the size and energy of the electron orbital. For the representative elements the principal quantum number for electrons in the outer most shell is given by the period in the periodic table. The principal quantum number for the transition metals lags one shell behind the period, and for the lanthanides and actinides lags two shells behind the period.

Valence electrons, the electrons which contribute most to an element's chemical properties, are located in the outer most shell of an atom. Typically, but not always, only electrons from the *s* and *p* subshells (explained below) are considered valence electrons.

The second quantum number is the **azimuthal quantum number, ℓ**. The azimuthal quantum number designates the **subshell**. These are the orbital shapes with which we are familiar such as ***s, p, d,* and *f***. If $\ell = 0$, we are in the *s* subshell; if $\ell = 1$, we are in the *p* subshell; and so on. For each new shell, there exists an additional subshell with the azimuthal quantum number **$\ell = n-1$**. Each subshell has a peculiar shape to its orbitals. You should recognize the shapes of the orbitals in the *s* and *p* subshells. These orbitals are actually probability functions of the position of the electron. By definition, there is a 90% chance of finding the electron somewhere on the surface of its orbital.

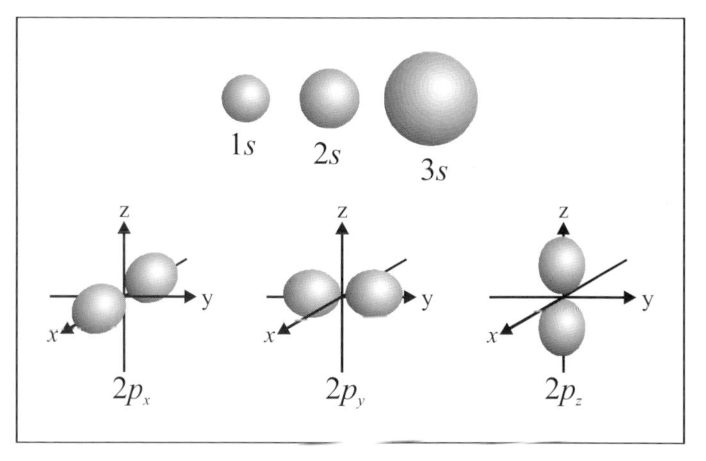

The third quantum number is the **magnetic quantum number, m_ℓ**. The magnetic quantum number designates the precise **orbital** of a given subshell. Each subshell will have orbitals with magnetic quantum numbers from **$-\ell$ to $+\ell$**. Thus for the first shell with $n = 1$, and $\ell = 0$, there is only one possible orbital, and its magnetic quantum number is 0. For the third shell with $n = 3$, and $\ell = 2$, there are 5 possible orbitals which have the magnetic quantum numbers of -2, -1, 0, $+1$, and $+2$.

The fourth quantum number is the **electron spin quantum number, m_s**. The electron spin quantum number can have values of **–½ or +½**. Any orbital can hold up to two electrons and no more. If two electrons occupy the same orbital, they have the same first three quantum numbers. **The Pauli exclusion principle** says that no two electrons in the same atom, can have the same four quantum numbers. Because two electrons in the same orbital have identical 1st, 2nd, and 3rd quantum numbers, they must have opposite electron spin quantum numbers.

The Heisenberg Uncertainty Principle

The Heisenberg Uncertainty Principle arises from the dual nature (wave-particle) of matter. It states that there exists an inherent uncertainty in the product of the position of a particle and its momentum, and that this uncertainty is on the order of Planck's constant.

$$\Delta x \Delta p \approx h$$

A common misunderstanding of the uncertainty principle is that its validity depends upon the interference of a measurement device, such as light. In other words, if we could somehow take measurements without disturbing what we are measuring, then the uncertainty principle would 'go away'. On the contrary, if we could take such measurements, we would find the same inherent uncertainty described by the uncertainty principle. The uncertainty principle stands as a fundamental principle of physics without regard to measurement device or technique.

Energy Level of Electrons

Nature typically prefers a lower energy state. The lower the energy level of a situation, the more stable the situation. Thus, electrons generally find an available orbital with the lowest energy state whenever they add to an atom. This is called the **Aufbau principle**. (Aufbau is German for 'build up'.) The orbital with the lowest energy will be contained in the subshell with the lowest energy.

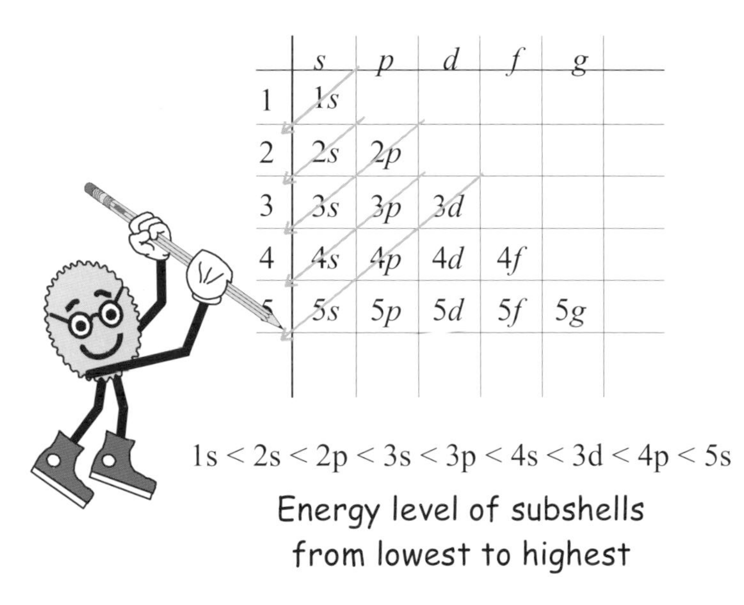

Energy level of subshells from lowest to highest

A simple trick to find the relative energies of the subshells is to use this table. The chart grows like stair-steps. An arrow is drawn down each diagonal as shown. If we follow the arrows as they go down the steps, they show us the order of increasing energy for the subshells. Notice that the energy levels are not exactly in numerical order. For example, the 4*s* subshell is at a lower energy level than the 3*d*.

To see why the energy level rises as the electrons move further from the nucleus, we must consider the attractive force between the negatively charged electrons and the positively charged nucleus. Because the force is attractive, we must do work to separate them; we apply a force over a distance. Work is the transfer of energy into or out of a system. In this case, our system is the electron and the nucleus. We are doing work on the system, so we are transferring energy into the system. This energy shows up as increased electrostatic potential energy. For reasons that we shall discuss in Physics Lecture 7, this energy increases from a large negative number to zero as the electron moves to an infinite distance away from the nucleus.

In some first row transition metals, the 4*s* and 3*d* orbitals are *degenerate* (they have nearly the same energy level). This means that sometimes electrons will add to an empty 3*d* orbital before filling the 4*s* orbital. In first row transition ions, the 4*s* orbitals are always at a higher energy level, and thus electrons will fill the 3*d* orbitals before the 4*s* orbitals when forming these ions.

You don't have to memorize the electron configuration of each transition metal. Just be aware that they don't always follow the table given above, due to degenerate orbitals.

	$1s$	$2s$	$2p_x$	$2p_y$	$2p_z$
hydrogen	↑				
helium	↑↓				
lithium	↑↓	↑			
beryllium	↑↓	↑↓			
boron	↑↓	↑↓	↑		
carbon	↑↓	↑↓	↑	↑	
nitrogen	↑↓	↑↓	↑	↑	↑
oxygen	↑↓	↑↓	↑↓	↑	↑

If we considered the energy of two particles with like charges, we would find that as the particles near each other, their potential energy increases. This is the case when electrons approach each other. It explains why only two electrons can fit into one orbital. It also helps explain **Hund's rule**: electrons will not fill any orbital in the same subshell until all orbitals in that subshell contain at least one electron. Hund's rule can be represented graphically as shown in the chart on the left. Electrons are represented by vertical arrows. Upward arrows represent electrons with positive spin, and downward arrows represent electrons with negative spin.

Max Planck is considered to be the father of quantum mechanics. **Planck's quantum theory** demonstrates that electromagnetic energy is quantized (i.e. it comes only in discreet units related to the wave frequency). In other words, if we transfer energy from one point to another via an electromagnetic wave, and we wish to increase the amount of energy that we are transferring without changing the frequency, we can only change the energy in discreet increments given by:

$$\Delta E = hf$$

where h is Planck's constant (6.6×10^{-34} J s).

Einstein showed that if we considered light as a particle phenomenon with each particle as a photon, the energy of a single photon is given by the same equation: $\underline{E_{photon} = hf}$. Neils Bohr unsuccessfully attempted to apply this theory to the energy of electrons using classical mechanics. Although Bohr's theory worked to explain the molecular orbitals of hydrogen, it failed with all other atoms. **Louis de Broglie** then showed that electrons and other moving masses exhibit wave characteristics that follow the equation:

$$\lambda = \frac{h}{mv}$$

The energy levels of electrons are quantized as well. The possible energy levels of an electron can be represented as an energy ladder. Each energy level is analogous to a rung on a ladder. The electrons may occupy any rung, but can never occupy the space between rungs because this space represents forbidden energy levels.

When an electron falls from a higher energy rung to a lower energy rung, energy is released from the atom in the form of a photon. The photon must have a frequency which corresponds to the change in energy of the electron as per $\Delta E = hf$. Of course the reverse is also true. If a photon collides with an electron, it can only bump that electron to another energy level rung and not between energy level rungs. If the photon doesn't have enough energy to bump the electron to the next rung, the electron will not move from its present rung and the photon will be reflected away.

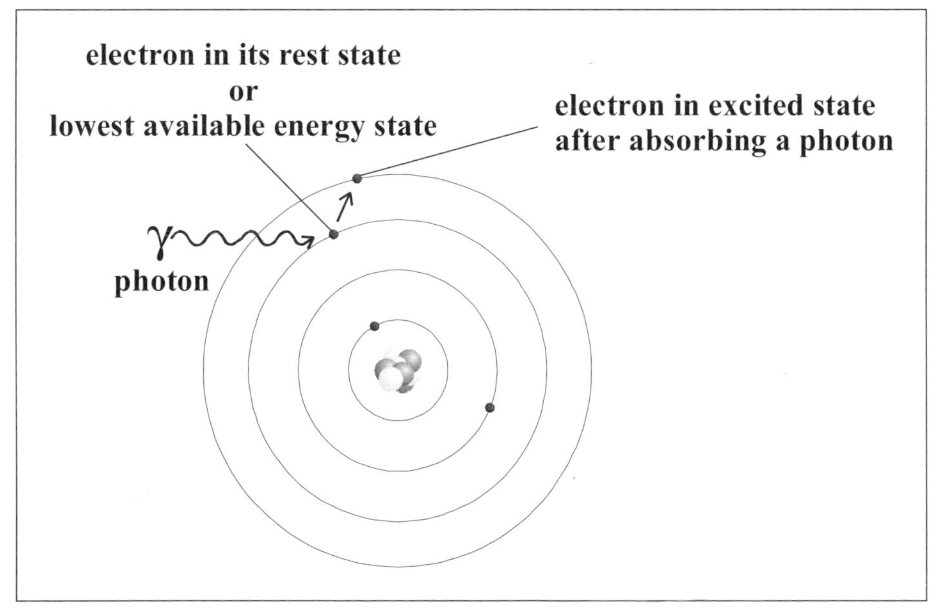

With the **photoelectric effect** Einstein demonstrated this one-to-one, photon to electron collision and that light was made up of particles. Einstein's reasoning went as follows: Light shining on a metal can cause the emission of electrons (the photoelectric effect). Since the energy of a wave is proportional to its intensity, we would expect that when the intensity of light shining on a metal is increased (i.e. more photons are used), the kinetic energy of an emitted electron would increase accordingly. This is not the case. Instead, the kinetic energy of the electrons increases only when the frequency increases. If the frequency is low enough, no electrons at all will be emitted regardless of the intensity. This demonstrates that the electrons must be ejected by one-to-one photon-electron collisions and not by the combined energies of two or more photons. It also shows that if a single photon does not have sufficient energy, no electron will be emitted. The minimum amount of energy required to eject an electron is called the *work function*, Φ, of the metal. The kinetic energy of the ejected electron is given by the energy of the photon minus the work function ($K.E. = hf - \Phi$).

Electron configurations list the electrons for a given atom, typically in the order of the subshell energy level as predicted by the Aufbau principle. To write an electron configuration, simply count the number of electrons in the atom, and then in each subshell. As a superscript to the right of each subshell, we write the number of electrons in that subshell. The total number of electrons in the configuration must equal the total number of electrons in the atom. Electron configurations for several atoms are given below:

$$\text{Na} \Rightarrow 1s^2\ 2s^2\ 2p^6\ 3s^1$$
$$\text{Ar} \Rightarrow 1s^2\ 2s^2\ 2p^6\ 3s^2\ 3p^6$$
$$\text{Fe} \Rightarrow 1s^2\ 2s^2\ 2p^6\ 3s^2\ 3p^6\ 4s^2\ 3d^6$$
$$\text{Br} \Rightarrow 1s^2\ 2s^2\ 2p^6\ 3s^2\ 3p^6\ 4s^2\ 3d^{10}\ 4p^5$$

An abbreviated electron configuration can be written by using the configuration of the next smallest noble gas as follows:

$$\text{Na} \Rightarrow [\text{Ne}]\ 3s^1$$
$$\text{Ar} \Rightarrow [\text{Ar}]$$
$$\text{Fe} \Rightarrow [\text{Ar}]\ 4s^2\ 3d^6\ \text{(sometimes written [Ar]}\ 3d^6\ 4s^2\text{)}$$
$$\text{Br} \Rightarrow [\text{Ar}]\ 4s^2\ 3d^{10}\ 4p^5$$

Above are the electron configurations for atoms whose electrons are all at their lowest energy levels. This is called the **ground state**. Electron configurations can also be given for ions, and atoms with excited electrons:

$$\text{Na}^+ \Rightarrow 1s^2\ 2s^2\ 2p^6\ \underline{\text{or}}\ [\text{Ne}]$$
$$\text{Fe}^{3+} \Rightarrow [\text{Ar}]\ 3d^5$$
$$\text{Br}^- \Rightarrow [\text{Ar}]\ 4s^2\ 3d^{10}\ 4p^6\ \underline{\text{or}}\ [\text{Kr}]$$
$$\text{Be}_{\text{with an excited electron}} \Rightarrow 1s^2\ 2s^1\ 2p^1$$

Notice that ions like to have an electron configuration like that of a noble gas. Also, notice that an electron can momentarily (for a matter of microseconds) absorb energy and jump to a higher energy level creating an atom in an excited state.

Notice the configuration of Fe^{3+} above. Cations always form by losing electrons from the subshells with the highest principal quantum number.

Questions 11 through 15 are **NOT** based on a descriptive passage.

11. Which of the following species has an unpaired electron in its ground-state electronic configuration?

A. Ne
B. Ca^+
C. Na^+
D. O^{2-}

12. What is the electron configuration of chromium?

A. [Ar] $3d^6$
B. [Ar] $4s^1\ 3d^5$
C. [Ar] $4s^2\ 3d^3$
D. [Ar] $4s^2\ 4p^4$

13. Which of the following best explains why sulfur can make more bonds than oxygen?

A. Sulfur is more electronegative than oxygen.
B. Oxygen is more electronegative than sulfur.
C. Sulfur has 3*d* orbitals not available to oxygen.
D. Sulfur has fewer valence electrons.

14. If the position of an electron is known with 100% certainty, which of the following cannot be determined for the same electron?

A. mass
B. velocity
C. charge
D. spin quantum number

15. When an electron moves from a 2*p* to a 3*s* orbital, the atom containing that electron:

A. becomes a new isotope.
B. becomes a new element.
C. absorbs energy.
D. releases energy.

Answers to Questions 11-15

11. B is correct. The quickest way to see this is by realizing that atoms like to form ions with electron configurations similar to the nearest noble gas. Of course a noble gas does not have any unpaired electrons. You should recognize that Ca likes to form a 2+ ion not a 1+ ion.

12. B is correct. This question borders on requiring too much specific knowledge for the MCAT. The knowledge that the 4*s* and 3*d* orbitals are at the same energy level for the first row transition metals is probably beyond the MCAT. Rather than memorize specific exceptions to the Aufbau principle, answer this question by eliminating that A, C, and D must be wrong. A is wrong because there is no reason to skip the *s* subshell entirely. C is wrong because it contains the wrong number of electrons. D is wrong because we should be in the 3*d* subshell, not the 4*d* subshell. You may be able to see that, by Hund's rule, each electron would rather take its own orbital than share an orbital at the same energy level with another electron. Thus for Chromium, electrons fill the orbitals like this:

$$\text{Chromium} \quad [\text{Ar}]\ \underset{4s}{\underline{\upharpoonleft}} \quad \underset{3d}{\underline{\upharpoonleft}}\ \underset{3d}{\underline{\upharpoonleft}}\ \underset{3d}{\underline{\upharpoonleft}}\ \underset{3d}{\underline{\upharpoonleft}}\ \underset{3d}{\underline{\upharpoonleft}}$$

instead of like this:

$$\text{Chromium} \quad [\text{Ar}]\ \underset{4s}{\underline{\upharpoonleft\downharpoonright}} \quad \underset{3d}{\underline{\upharpoonleft}}\ \underset{3d}{\underline{\upharpoonleft}}\ \underset{3d}{\underline{\upharpoonleft}}\ \underset{3d}{\underline{\upharpoonleft}}\ \underset{3d}{\underline{\quad}}$$

Copper is the only other first row transition metal that breaks the Aufbau principle. Its electron configuration is [Ar] $4s^1\ 3d^{10}$. Also remember that for first row transition metal ions, the 3*d* orbitals are at a lower energy level than the 4*s* orbitals, so the 4*s* and not the 3*d* electrons tend to be removed to form the ions in these metals.

13. C is correct. Because sulfur is larger than oxygen, sulfur has 3*d* subshells available that allow electrons to form bonds and break the octet rule of the Lewis structure.

14. B is correct. According to the Heisenberg uncertainty principle, both the position and the momentum *mv* of an electron cannot be known with absolute certainty at the same time. Since we know the mass of an electron, the uncertainty must lie in the velocity.

15. C is correct. The atom must absorb energy in order for one of its electrons to move to a higher energy level orbital.

Lecture 2: Gases, Kinetics, and Chemical Equilibrium

Gases

To better understand the complex behavior of gases, scientists have theorized a model of an **ideal gas**. This model is called the **kinetic molecular theory**. In the kinetic molecular theory, an ideal gas lacks certain real gas characteristics. Ideal gas differs from real gas in the following four ways:

1) gas molecules have zero volume;
2) gas molecules exert no attractive forces on each other;
3) collisions between gas molecules are completely elastic;
4) the average kinetic energy of gas molecules is directly proportional to the temperature of the gas.

Ideal gas obeys **the ideal gas law**:

$$PV = nRT$$

where P is the pressure in atmospheres, V is the volume in liters, n is the number of moles of gas, R is the universal gas constant (0.08206 L atm K^{-1} mol^{-1}, or 8.314 J K^{-1} mol^{-1}), and T is the temperature in kelvins. The ideal gas law is derived from **Charles' law** (the volume of a gas is proportional to temperature at constant pressure), **Boyle's law** (the volume of a gas is inversely proportional to pressure at constant temperature), and **Avogadro's law** (the volume of a gas is proportional to the number of moles at constant temperature and pressure).

Rather than memorizing Charles', Boyle's, and Avogadro's laws, you should have a good understanding of the ideal gas law. The equations often associated with these laws have been left out purposely, so that you won't memorize them. PV = nRT will solve any problem that the other three laws might solve. And, you definitely won't need to know each law by its name.

Notice that the ideal gas law does not change for different gases. (Of course not. It's written for an ideal gas.) This means that all gases (behaving ideally) will have the same volume, if they have the same temperature, pressure, and number of molecules. 0 °C and 1 atm is called **standard temperature and pressure (STP).** At STP one mole of any gas (behaving ideally) will occupy the **standard molar volume of 22.4 liters**.

You must recognize the standard molar volume and understand what it means. Learn to use this volume with the ideal gas law. For instance, 2 moles of gas at 0 °C occupying 11.2 liters will have a pressure of 4 atm. To get this result, we start with the standard molar volume, 22.4 L, at STP. First, we double the moles, so, according to the ideal gas law, the pressure doubles. Second, we halve the volume, so pressure doubles again.

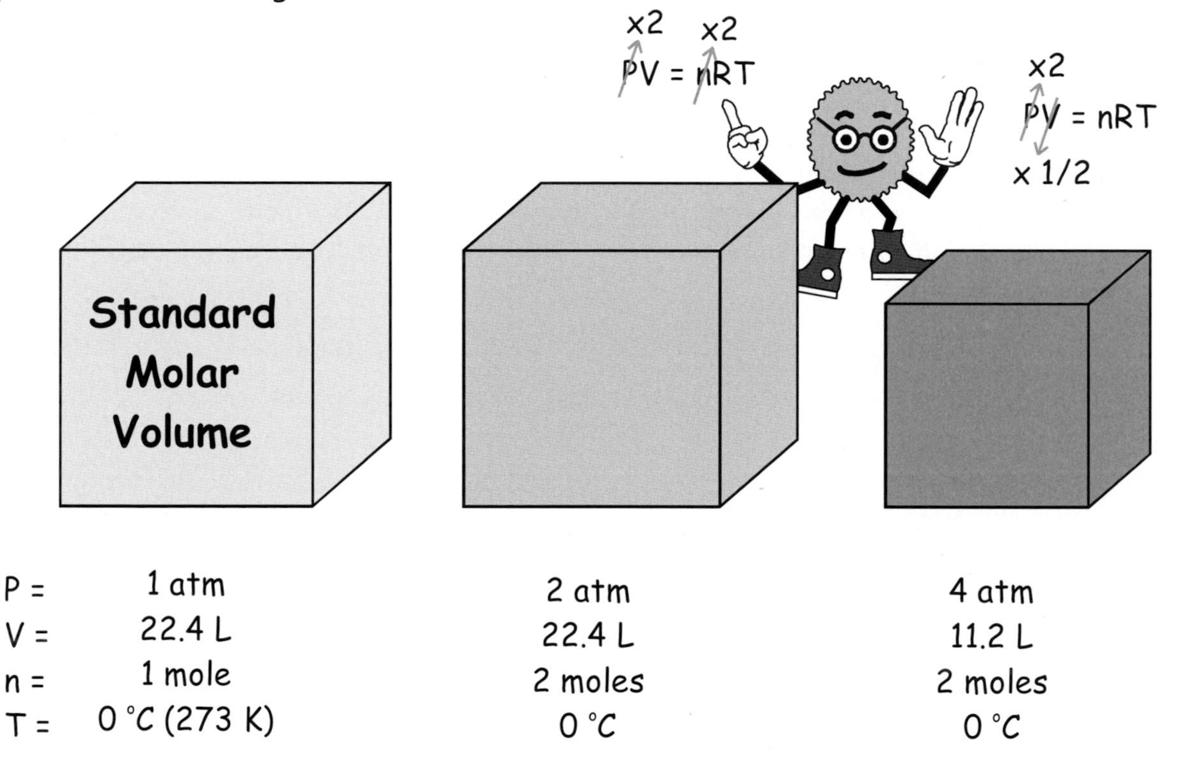

If we take this idea still further, we can say that all mixtures of gases (behaving ideally) will have the same pressure, if they have the same temperature, volume, and number of molecules. This must mean that in a mixture of gases, each gas contributes to the pressure in the same proportion as it contributes to the number of molecules of the gas. This makes sense, given the kinetic molecular theory because molecules have no volume, no interactive forces, and kinetic energy is conserved when they collide; thus, each gas essentially behaves as if it were in the container by itself. The amount of pressure contributed by a single gas in a gaseous mixture is called the **partial pressure** of that gas. **Dalton's law** states that the total pressure exerted by a gaseous mixture is the sum of the partial pressures of each of its gases. This is just another way of saying that each gas makes the same contribution to the total pressure as it would if it were alone in the container. This amount is the partial pressure for the respective gas. The equation for the partial pressure is:

$$P_a = \chi_a P_{total}$$

where P_a is the partial pressure of gas 'a', and χ_a is the mole fraction of gas 'a'. (The mole fraction is the number of moles of gas 'a' divided by the total number of moles of gas in the sample...Chemistry Lecture 4).

From the ideal gas law, we can derive the following equation relating average kinetic energy and the temperature of a gas:

$$K.E._{avg} = \frac{3}{2}RT$$

where the average kinetic energy is found from the rms velocity. (rms velocity is the square root of the average of the squares of the molecular velocities.) Notice that this is the average kinetic energy (and not the energy of all, or maybe even any) of the molecules. A gaseous molecule chosen at random may have almost any kinetic energy associated with it.

Since the temperature dictates the average kinetic energy of the molecules in a gas, the gas molecules of each gas in any gaseous mixture must have the same average kinetic energy. For instance, the air we breath is made up of approximately 21% O_2 and 79% N_2. Thus, the molecules of O_2 and N_2 gas have the same average kinetic energy. However, this means that they have different average velocities. By setting their kinetic energies equal to each other, we can derive a relationship between their velocities. This relationship, which gives the ratio of the rms velocities of two gases in a homogeneous mixture, is called **Graham's law**:

$$\frac{v_1}{v_2} = \frac{\sqrt{m_2}}{\sqrt{m_1}}$$

Notice that the subscripts are inverted from one side of the equation to the other. Graham's law also tells us that the velocity of the molecules of a pure gas is inversely proportional to the square root of its mass.

Interestingly, Graham's law gives information about the rates of two types of gaseous spreading: **effusion** and **diffusion**. Effusion is the spreading of a gas from high pressure to low pressure through a 'pin hole'. (A 'pin hole' is defined as an opening much smaller than the average distance between the gas molecules.) Because molecules of a gas with higher rms velocity will experience more collisions with the walls of a container, the rate at which molecules from such a gas find the pin hole and go through is likely to be greater. In fact, Graham's law predicts the comparative rates of effusion for two gases at the same temperature. The comparative rates of effusion of two gases is equal to the inverse of the ratio of the square roots of their molecular weights and equal to the ratio of their rms velocities.

$$\frac{\text{effusion rate}_1}{\text{effusion rate}_2} = \frac{\sqrt{M_2}}{\sqrt{M_1}}$$

Diffusion is the spreading of one gas into another at equal pressure. The ratio of the diffusion rates of two gases (acting ideally) is given by Graham's law. However, the diffusion rate is much slower than the rms velocity of the molecules. The reason for this is that gas molecules collide with each other and with molecules of other gases as they diffuse. For instance, if we wet two cotton balls, one with aqueous NH_3 and the other with aqueous HCl, and place them in opposite ends of a glass tube, gaseous NH_3 and HCl will diffuse toward each other through the air inside the tube. Where they meet, they will react to form NH_4Cl, which will precipitate as a white solid. Graham's law accurately predicts that NH_3 will travel 1.5 times further than HCl. However, any particular molecule is likely to follow a very crooked path similar to those shown in the diagram.

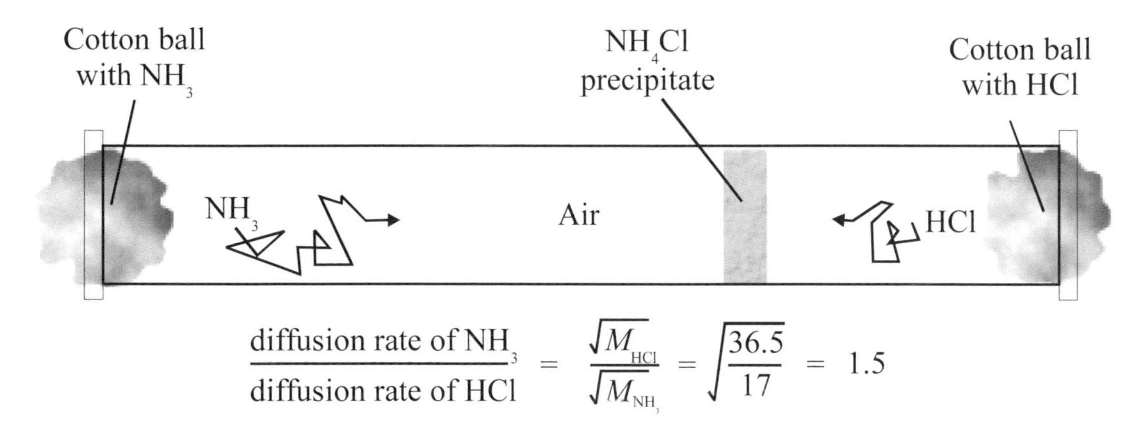

$$\frac{\text{diffusion rate of NH}_3}{\text{diffusion rate of HCl}} = \frac{\sqrt{M_{HCl}}}{\sqrt{M_{NH_3}}} = \sqrt{\frac{36.5}{17}} = 1.5$$

Real Gases

Of course, no real gas exhibits perfect ideal behavior; however, at pressures of one atmosphere or less, substances that we typically think of as gasses exhibit ideal behavior. **Gasses deviate from ideal behavior at low temperature and high pressure**.

You should be aware of how real gases deviate from ideal behavior. *Van der Waals equation*:

$$[P + a(n/V)^2](V - nb) = nRT,$$

gives the real pressure and real volume of a gas, where a and b are constants for specific gases. However, you do not need this equation for the MCAT and, in fact, it is far more important that you have a qualitative understanding of real gas deviations from ideal behavior. First, since molecules of a real gas do have volume, their volume must be added to the ideal volume. Thus:

$$\boldsymbol{V_{\text{real}} > V_{\text{ideal}}}$$

where V_{ideal} is calculated from $PV = nRT$.

Second, since molecules in a real gas do exhibit forces on each other, and those force are attractive when the molecules are far apart, molecules slow before colliding with container walls, and strike with less force than predicted by the kinetic molecular theory. Thus a real gas exerts less pressure than predicted by the kinetic molecular theory.

$$\boldsymbol{P_{\text{real}} < P_{\text{ideal}}}$$

where P_{ideal} is calculated from $PV = nRT$.

Questions 16 through 20 are **NOT** based on a descriptive passage.

16. A 13 gram gaseous sample of an unknown hydrocarbon occupies a volume of 11.2 L at STP. What is the hydrocarbon?

A. CH
B. C_2H_4
C. C_2H_2
D. C_3H_3

17. If the density of a gas is given as ρ, which of the following expressions represents the molecular weight of the gas?

A. $P/\rho RT$
B. $\rho RT/P$
C. $nRT/P\rho$
D. $P\rho/RT$

18. Ammonia burns in air to form nitrogen dioxide and water.

$$4NH_3(g) + 7O_2(g) \rightarrow 4NO_2(g) + 6H_2O(l)$$

If 8 moles of ammonia are reacted with 14 moles of oxygen gas in a rigid container with an initial pressure of 11 atm, what is the partial pressure of nitrogen dioxide in the container when the reaction runs to completion? (Assume constant temperature.)

A. 4 atm
B. 6 atm
C. 11 atm
D. 12 atm

19. At moderately high pressures, the PV/RT ratio for one mole of methane gas is less than one. The most likely reason for this is:

A. Methane gas behaves ideally at moderate pressures.
B. The volume of methane gas molecules creates a greater deviation from ideal gas behavior than the intermolecular forces between methane gas molecules.
C. The intermolecular forces between methane gas molecules create a greater deviation from ideal gas behavior than the volume of methane gas molecules.
D. The temperature must be very low.

20. Equal molar quantities of oxygen and hydrogen gas were placed in container A under high pressure. A small portion of the mixture was allowed to effuse for a very short time into the vacuum in container B. Which of the following is true concerning partial pressures of the gases at the end of the experiment?

A. The partial pressure of hydrogen in container A is approximately four times as great as the partial pressure of oxygen in container A.
B. The partial pressure of oxygen in container A is approximately four times as great as the partial pressure of hydrogen in container A.
C. The partial pressure of hydrogen in container B is approximately four times as great as the partial pressure of oxygen in container B.
D. The partial pressure of oxygen in container B is approximately four times as great as the partial pressure of hydrogen in container B.

Answers to Questions 16-20

16. **C is correct.** You should recognize that 1 mole of gas occupies 22.4 liters at STP, so there is 0.5 moles of gas in the sample. 13 g/0.5 mol = 26 g/mol.

17. **B is correct.** Since density (ρ) is mass (m) divided by volume (V), and mass is moles (n) times molecular weight (MW), we have $(n\text{MW})/V = \rho$. After some algebra we have: MW = $(\rho V)/n$. From the ideal gas law we know that $V/n = RT/P$. Substituting we have answer B.

18. **A is correct.** The number of moles of gas is extra information. If the container began at 11 atm then each gas is contributing a pressure in accordance with its stoichiometric coefficient. When the reaction runs to completion, the only gas in the container is nitrogen dioxide, so the partial pressure of nitrogen dioxide is the total pressure. The volume of the container remains constant, so the pressure in accordance with the stoichiometric coefficient of nitrogen dioxide.

19. **C is correct.** An ideal gas has a PV/RT equal to one. Real volume is greater than predicted by the kinetic molecular theory, and real pressure is less than predicted by the kinetic molecular theory. Volume deviations are due to the volume of the molecules, and pressure deviations are due to the intermolecular forces. Thus, a negative deviation in this ratio would indicate that the intermolecular forces are having a greater affect on the nonideal behavior than the volume of the molecules.

20. **C is correct.** From Grahams law we know that the effusion rate for hydrogen is four times that of oxygen.

$$\frac{H_{2\ \text{effusion rate}}}{O_{2\ \text{effusion rate}}} = \sqrt{\frac{\text{M.W.}_{\text{oxygen}}}{\text{M.W.}_{\text{hydrogen}}}} = \sqrt{\frac{32}{2}} = 4$$

Since we don't know how many moles of gas were initially in container A, nor how many moles effused out, we don't know the ratio of hydrogen to oxygen. However, since we know that four times as many moles of hydrogen effused from container A into B, we know that container B contains four times as many moles of hydrogen. We can neglect any effusion in the reverse direction since the question says a "very short time".

Kinetics

Chemical kinetics deals with the rate at which a reaction occurs. The **initial rate** of a reaction is defined as the change in the concentration of the reactants per unit of time. Thus the units of rate are mol L^{-1} s^{-1}. Since chemical reactions are reversible, the reverse reaction increases the concentration of reactants and complicates the study of kinetics. To avoid this complication, problems often deal only with reaction rates where the concentration of products is so small as to have a negligible effect on the rate. We can do this by removing the products from a reaction or by only considering the initial rate of a reaction before significant concentrations of product have formed. Initial rates depend only upon the concentration of reactants and are independent of the concentration of products.

It is important to understand that the rate at which a reaction occurs and the tendency for that same reaction to occur are not the same thing. For instance, hydrogen and oxygen gas combine to form liquid water releasing large amounts of energy. However, at 25 °C the reaction is so slow that they can exist together indefinitely. The reaction is spontaneous (Lecture 3 will discuss spontaneity) but its spontaneity indicates nothing about the rate at which it occurs.

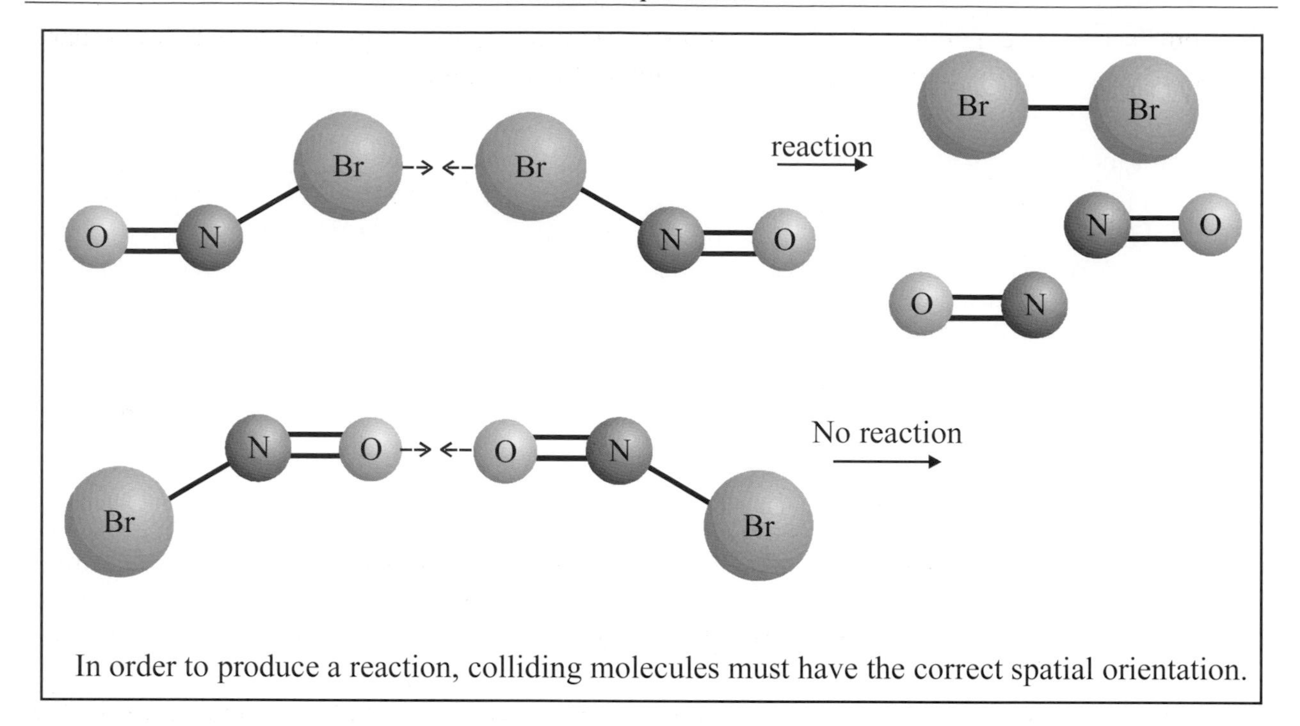

In order to produce a reaction, colliding molecules must have the correct spatial orientation.

The *collision model* of reactions provides an enlightening method for visualizing chemical reactions. In order for a chemical reaction to occur, the reacting molecules must collide. However, the rate of a given reaction is found to be much smaller than the frequency of collisions. This indicates that not all collisions result in new molecules. There are two requirements for a given collision to result in the rearrangement of atoms and the formation of new compounds. First, the combined kinetic energies of the colliding molecules must reach a threshold energy called the **activation energy**. Second, the colliding molecules must have the proper spatial orientation. We can combine into an equation the collision frequency z, the fraction of these collisions having the effective spatial orientations p (called the *steric factor*), and the fraction of these collisions having sufficient energy $e^{-Ea/RT}$ (where E_a is the activation energy). The result is the *Arrhenius equation*:

$$k = zpe^{-Ea/RT}$$

(often written as $k = Ae^{-Ea/RT}$). k is called the rate constant. As demonstrated by the rate law (discussed later in this lecture), the rate constant is directly proportional to the rate of a reaction.

Notice from the Arrhenius equation that, as the temperature T increases, the activation energy E_a remains constant, but that $e^{-Ea/RT}$, the fraction of collisions that have at least the activation energy, increases. Thus the rate constant k increases with increasing temperature for all reactions. It follows that the rate of a reaction also increases with temperature. Remember, do not confuse an increase in the rate with an increase in the tendency of the reaction to occur.

The temperature dependence of rate can be demonstrated by the graph below, which compares two samples of the same gaseous mixture at different temperatures.

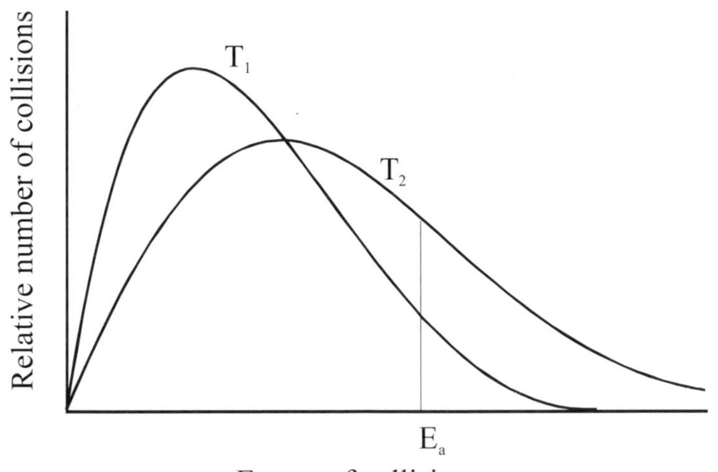

The area under any section of the curve represents the relative number of collisions in that energy range. Notice the difference between the areas under the curves to the right of the energy of activation. Notice also that the energy of activation is the same at both temperatures.

The rate of a reaction can be described mathematically by its **rate law**. A sample rate law for the hypothetical chemical reaction $A + B \rightarrow C$ is:

$$\textbf{Rate} = k[A]^x[B]^y$$

where x and y are the **order of each respective reactant** and the sum x+y is the **overall order** of the reaction. [A] and [B] are the concentrations of the reactants. (By convention, the square brackets [ ] indicate concentrations.) Both the order of the reactants and the rate constant must be established through experiment. The above equation assumes that only negligible amounts of product are involved in the reaction.

Any reaction can be separated into **elementary steps**. An elementary step indicates the number of species that must collide in order for a reaction to occur (also called *molecularity*). The rate of the slowest elementary step determines the rate of the overall reaction and is called the **rate determining step**. As an example we can look at the following reaction:

$$NO_2(g) + CO(g) \rightarrow NO(g) + CO_2(g)$$

This reaction has two elementary steps:

1) $NO_2(g) + NO_2(g) \rightarrow NO_3(g) + NO(g)$ *slow step*

2) $NO_3(g) + CO(g) \rightarrow NO_2(g) + CO_2(g)$ *fast step*

Notice that if we add these two equations together, we arrive at the original equation. Since the first step is the slow step, the rate law for the overall reaction is given by this step and is:

$$\text{rate} = k[NO_2]^2$$

We know that the exponent is 2 because we derived the rate law from an elementary equation, and in the elementary equation two NO_2 molecules reacted together. This reasoning only works for elementary equations. The only way to know if an equation is elementary is if you are told that it is elementary. Don't forget, the above rate law assumes negligible contribution from the reverse reaction, and it also assumes a sufficient concentration of CO for the fast step to occur.

You must recognize that the slow step determines the rate. It's not actually the step that takes time. The step itself is just a collision, which requires a small fraction of a second for any reaction. The time is spent waiting for the collision to happen. In other words, the slow step is a collision which has less probability of occurring than the fast step, and thus doesn't happen as often. For instance, I start with three boxes, the first two representing a step in a reaction and the third box to hold the products. I begin with 1000 marbles (the reactants) in the first box. Now I remove one marble from the first box each second and place it in the second box, and I remove one from the second box every 2 seconds and place it in the third box. What is the average rate at which reactants are changing to products, or at what rate do I have marbles moving from the first box to the third. After 200 seconds, I have 200 marbles moving to the second box but only 99 marbles have completed the reaction. (99 and not 100 because one second was required before any marbles were in the second box.) Thus the rate is 99 marbles/200 seconds, or approximately 1 marble every two seconds; the rate of the slow step.

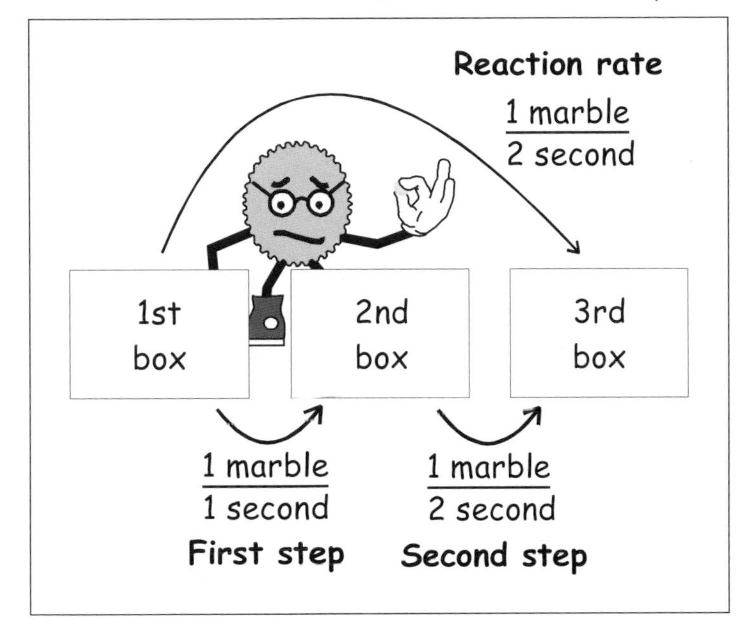

If you are not given the elementary equation for a reaction, you must find the order of the reactants by experiment. If we imagine the hypothetical reaction:

$$2A + B + 3C \rightarrow D + E$$

and we are given the following table with experimental data:

Experiment	Initial concentration of A (mol/L)	Initial concentration of B (mol/L)	Initial concentration of C (mol/L)	Measured Initial Rate (mol/L s)
1	0.1	0.1	0.1	8.0×10^{-4}
2	0.2	0.1	0.1	1.6×10^{-3}
3	0.2	0.2	0.1	6.4×10^{-3}
4	0.1	0.1	0.2	8.0×10^{-4}

we can find the order of each reactant by comparing the rates between two experiments in which only the concentration of one of the reactants is changed. For instance, from experiment 1 to experiment 2, the initial concentration of A is doubled and the concentrations of the B and C remain the same. The rate also doubles. Thus the initial rate of this reaction is directly proportional to the concentration of A. Therefore, in the rate law, reactant A gets an exponent of 1, and the reaction is first order with respect to reactant A. Comparing experiments 2 and 3 we find that when only the concentration of B is doubled, the initial reaction is quadrupled. This indicates that the rate is proportional to the square of the concentration of B. Reactant B gets a 2 for its exponent in the rate law and the reactant is second order with respect to B. Finally, by comparing experiments 1 and 4 we see that nothing happens to the rate when the concentration of C is doubled. This means that the rate is independent of the concentration of C, and C gets an exponent of zero in the rate law. The complete rate law for our hypothetical reaction is:

$$\text{rate} = k[A]^1[B]^2[C]^0$$

also written as

$$\text{rate} = k[A][B]^2$$

By adding the exponents we find that the reaction overall is third order. Notice that the coefficients in the balanced chemical equation were not used to figure out the order. This is because we did not know whether or not we had an elementary reaction. By deriving the rate law from the experiments, we find that we did not.

Once we have derived the rate law from the experimental data, we can plug in the rate and concentrations from <u>any</u> of the experiments into the rate law, and solve for the rate constant *k*.

Notice that for an elementary step, the rate is increased by increasing the concentration of the reactants. If we consider the collision model, this makes sense. The greater the concentration of a species, the more likely are collisions between them, and the rate of a reaction is governed by probability of collisions.

A **catalyst** is a substance that increases the rate of a reaction without being consumed or permanently altered. Almost every chemical reaction in the human body is quickened by a protein catalyst called an **enzyme**. A catalyst increases the reaction rate by lowering the activation energy and increasing the fraction of effective collisions at a given temperature. Since the relative amount of effective collisions is greater, the probability is also greater, and the reaction rate is increased.

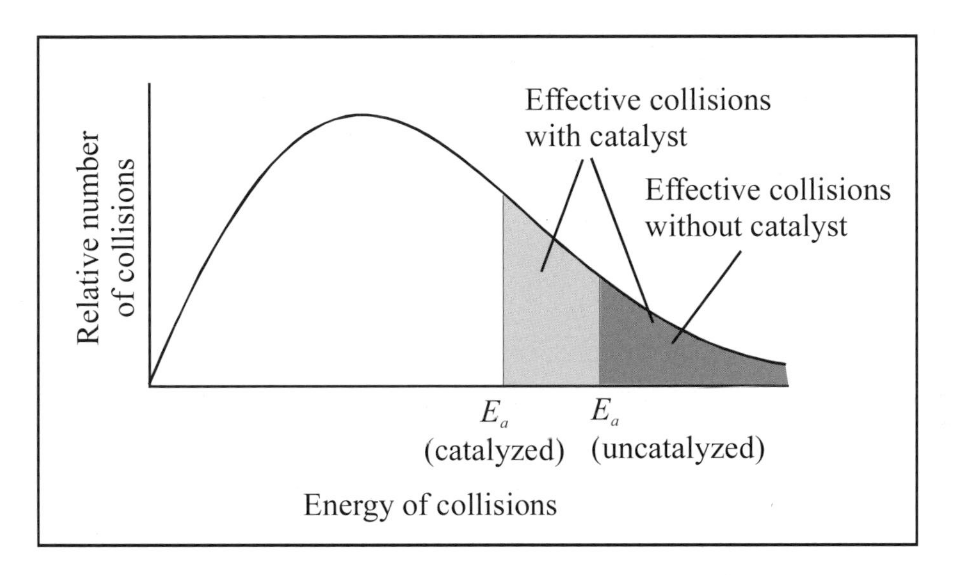

For the MCAT you must be able to derive the order of reactants from a table of experiments similar to the one given in this lecture. When given a rate law, You must be able to predict what changing the concentration of a reactant will do to the rate. Remember that a catalyst increases the rate of a reaction by lowering the activation energy of that reaction, and that a catalyst is never used up. Finally, remember that the rate of a reaction always increases with temperature, and you don't have to know the Arrhenius equation, but it should remind you that the rate constant must increase as temperature increases.

Questions 21 through 25 are **NOT** based on a descriptive passage.

21. Which of the following changes to a reaction will always increase the rate constant for that reaction?

A. decreasing the temperature
B. increasing the temperature
C. increasing the concentration of the reactants
D. increasing the concentration of the catalyst

22. All of the following may be true concerning catalysts and the reaction which they catalyze EXCEPT:

A. Catalysts are not used up by the reaction.
B. Catalysts lower the energy of activation.
C. Catalysts increase the rate of the reverse reaction.
D. Catalysts shift the reaction equilibrium to the right.

23. The table below shows 3 trials where the initial rate was measured for the reaction:

$$2A + B \rightarrow C$$

Which of the following expressions is the correct rate law for the reaction?

T	Molarity of A (mol/L)	Molarity of B (mol/L)	Initial Rate (mol/L/s)
1	0.05	0.05	5×10^{-3}
2	0.05	0.1	5×10^{-3}
3	0.1	0.05	1×10^{-2}

A. rate = 0.1[A]
B. rate = [A]
C. rate = [A][B]
D. rate = $[A]^2[B]$

24. The reaction below proceeds via the two step mechanism as shown.

Overall Reaction: $2NO_2 + F_2 \rightarrow 2NO_2F$

Step 1: $NO_2 + F_2 \rightarrow 2NO_2F + F$

Step 2: $NO_2 + F \rightarrow NO_2F$

X is the rate of step 1, and Y is the rate of step 2. If step 1 is much slower than step 2, then the rate of the overall reaction can be represented by:

A. X
B. Y
C. X + Y
D. X – Y

25. As temperature is increased in an exothermic gaseous reaction, all of the following increase EXCEPT:

A. reaction rate.
B. rate constant.
C. activation energy.
D. rms molecular velocity.

Answers to Questions 21-25

21. **B is correct.** Changing the concentration of the reactants will not change the rate constant. Increasing the concentration of a catalyst will only increase the rate of the reaction if the supply of catalyst is so small that the reactants are waiting for a catalyst. Most of the time on the MCAT, you can assume that the supply of catalyst is large enough so that a change in concentration will not change the reaction rate. (See Biology Lecture 2 for the graph relating reaction rate to enzyme catalysts.) Increasing the amount of catalyst never increases the rate constant. Increasing the temperature will always increase the rate constant, and the rate of the reaction. If the reaction is catalyzed by an enzyme, the enzyme may denature, slowing the reaction; however, the reaction without the enzyme is considered a different reaction.

22. **D is correct.** Catalysts do not directly affect the equilibrium of a reaction. Catalysts do increase the rate of the reverse reaction as well as the forward reaction.

23. **A is correct.** When the concentration of B is doubled, the rate doesn't change. When the concentration of A is doubled, the rate doubles. The reaction is first order overall, and first order with respect to A. By choosing a trial and plugging the values into the rate law, we find that the rate constant has a value of 0.1.

24. **A is correct.** The slow step determines the rate of a reaction.

25. **C is correct.** Exothermicity concerns the thermodynamics of the reaction, and not the rate. You can ignore it. The energy of activation is the energy required for a collision of properly oriented molecules to produce a reaction. This does not change with temperature.

Equilibrium

As the forward reaction proceeds, the concentration of reactants diminishes, and the rate of the forward reaction slows. If the reaction runs to completion, the concentration of at least one of the reactants will reach zero, and, of course, at that point the rate of the forward reaction must also reach zero. A reaction can run to completion when the products are removed during the reaction. However, since chemical reactions are reversible (see Chemistry Lecture 3 for reversibility of reactions), if the products are not removed, the reverse reaction begins creating reactants from products. If we begin a reaction with only reactants and no products, then initially there will be no reverse reaction, but as the concentration of the products builds, so will the rate of the reverse reaction. Thus, the rate of the forward reaction decreases, and the rate of the reverse reaction increases until the two rates become equal. This condition, where the forward reaction rate equals the reverse reaction rate, is called **chemical equilibrium**. At chemical equilibrium, there is no change in the concentration of the products or reactants. Notice that this conflicts with the definition of *initial rate*, which states that the rate of a reaction is given by the rate of change of the reactants. This is because the initial rate definition ignores the formation of reactants from the reverse reaction.

You must understand for the MCAT that equilibrium is dynamic. When a reaction reaches equilibrium, it does not mean that no reaction is taking place; it means that the forward and reverse reactions are taking place at the same rate. Although in a reaction at equilibrium there is no change in the concentration of reactants, the rates of the forward and reverse reactions are not zero.

Consider the reaction:

$$H_2O + CO \rightleftharpoons H_2 + CO_2$$

The graph below shows the reaction rates for the forward and reverse reactions over time, beginning with equal molar quantities of water and carbon monoxide.

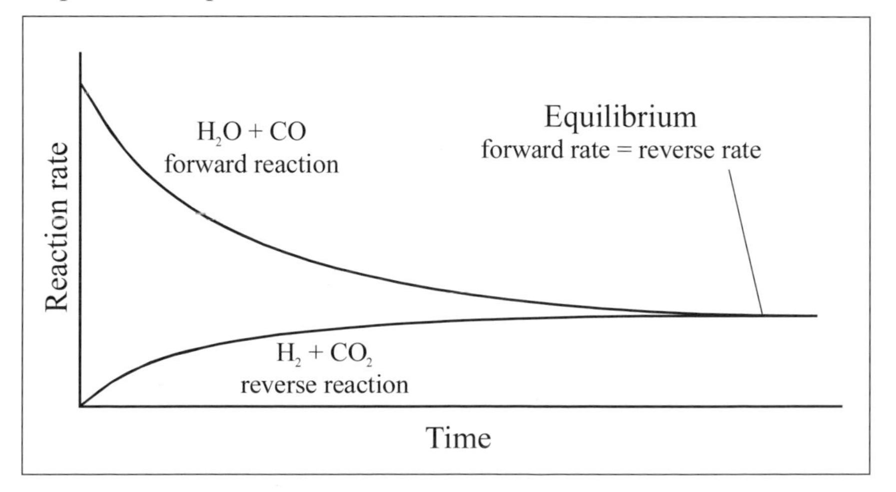

Notice that the reverse reaction rate begins at zero because initially there are no products. The forward reaction rate slows as the concentration of water and carbon monoxide diminishes. If we assume this is an elementary reaction, we can deduce that the rate constant for the forward reaction is greater than that for the reverse, and thus the equilibrium concentrations of the products are greater than those of the reactants. Look carefully at the graph and see if you agree. This is the type of graphical interpretation that is seen on the MCAT. Since the change in the number of molecules in the reactants and products is the same for each collision in this particular reaction, and the forward reaction rate changes faster than the reverse reaction rate, the rate constant for the forward reaction must be greater than for the reverse.

If we go back to our hypothetical reaction: $2A + B + 3C \rightarrow D + E$, and we imagine this time that the equation is elementary (unlikely because so many molecules are involved), then we have the following for the forward and reverse rate laws:

$$\text{Rate}_{\text{forward}} = k_f[A]^2[B][C]^3$$

$$\text{Rate}_{\text{reverse}} = k_r[D][E]$$

(**WARNING:** For the MCAT, never use the coefficients as the exponents in the rate law. We do it here because we are pretending that the equation is elementary.) Since equilibrium occurs when these two rates are equal, then, for equilibrium conditions only, we set them equal to each other as shown below:

$$k_f[A]^2[B][C]^3 = k_r[D][E]$$

With a little algebraic manipulation we have:

$$\frac{k_f}{k_r} = \frac{[D][E]}{[A]^2[B][C]^3}$$

Since both k's are constants, we replace them with a new constant called the **equilibrium constant K**. Notice that the equilibrium constant is a capital K. (**2nd WARNING:** This simple relationship between K equilibrium and k rate is only true for elementary equations. It is given here to deepen your insight into chemical reactions and not for use on the MCAT.) The relationship between a chemical equation and the equilibrium constant is called the **law of mass action** and is written as follows:

$$K = \frac{\textbf{Products}^{\text{coefficients}}}{\textbf{Reactants}^{\text{coefficients}}}$$

The law of mass action is good for all chemical equations, including non-elementary equations. In other words, for equilibrium constants, use the chemical equation coefficients as the exponents of the concentrations regardless of molecularity.

Remember, for equilibrium use the coefficients as exponents, and for rate don't use the coefficients as exponents.

There is one exception to using the coefficients as exponents in the law of mass action; pure liquids and pure solids always receive an exponent of zero because their concentration does not change significantly, and thus does not affect the rate of the forward or reverse reactions. In other words, we just leave pure liquids and pure solids out of the equilibrium equation. For the MCAT, this usually means leaving out water and all solids. (We call them 'pure' because they are in a separate phase from the rest of the reactants. For instance, in a solution of salt water, water is in a liquid phase, while the salts are in an aqueous phase. See Chemistry Lecture 4 for more on phases.)

Don't use water or solids in the law of mass action.

The law of mass action is only true when a reaction has reached equilibrium. For reactions not at equilibrium we use a similar equation:

$$Q = \frac{\mathbf{Products}^{\text{coefficients}}}{\mathbf{Reactants}^{\text{coefficients}}}$$

where Q is called the **reaction quotient**. Of course, Q is not a constant; it can have any positive value. Q is simply a ratio of products over reactants at any time during the reaction.

Since reactions always move toward equilibrium, Q will always change toward K. Thus we can compare Q and K for a reaction at any given moment, and learn in which direction the reaction will proceed.

If Q is equal to K, then the reaction is at **equilibrium**.

If Q is greater than K, then the ratio of the concentration of products to the concentration of reactants, as given by the equation above, is greater than when at equilibrium, and the reaction increases reactants and decreases products. In other words, the reverse reaction rate will be greater than the forward rate. This is sometimes called a **leftward shift**.

If Q is less than K, then the ratio of the concentration of products to the concentration of reactants, as given by the equation above, is less than when at equilibrium, and the reaction increases products and decreases reactants. In other words, the forward reaction rate will be greater than the reverse rate. This is sometimes called a **rightward shift**.

For reactions involving gases, the equilibrium constant can be written in terms of partial pressures. The concentration equilibrium constant and the partial pressure equilibrium constant do not have the same value, but they are related by the equation:

$$K_p = K(RT)^{\Delta n}$$

where K_p is the partial pressure equilibrium constant and Δn is the sum of the coefficients of the products minus the sum of the coefficients of the reactants. This equation is not required for the MCAT, but you must be able to work with partial pressure equilibrium constants.

There is a general rule called **Le Chatelier's principle** that sometimes can be applied to systems at equilibrium. Le Chatelier's principle states that when a stress is applied to a system at equilibrium, the system will shift in a direction to reduce that stress.

There are three types of stresses that usually follow Le Chatelier's principle: 1) changing the concentration of a product or reactant; 2) changing the pressure of the system; 3) heating or cooling the system.

Le Chatelier's principle is very important for the MCAT. Let's examine the all gas reaction:

$$N_2 + 3H_2 \rightarrow 2NH_3$$

called the Haber process. The Haber process is an exothermic reaction, so it creates heat (see Chemistry Lecture 3). If we add N_2 or H_2 gas, Le Chatelier's principle predicts a shift to the right. The system attempts to compensate for the increased concentration of reactants by reducing their concentration with the forward reaction. Of course the reverse would be predicted, if we added NH_3 gas.

Since the forward reaction creates heat and the reverse reaction absorbs heat, Le Chatelier's principle predicts that the addition of heat would push the reaction to the left in order to absorb the extra heat.

Since the forward reaction creates 2 moles of gas and the reverse reaction creates 4 moles, if we add pressure, we expect a rightward shift in order to reduce the number of gas molecules and thus reduce the pressure.

Be careful! Sometime, especially in solvation reactions, Le Chatelier's principle does not predict the correct shift due to a temperature change.

It is also possible to increase the pressure of an all gas reaction such as the Haber process and not change the equilibrium.

Imagine a hypothetical all gas reaction similar to the Haber process:

$$A(g) + 3B(g) \rightleftharpoons 2C(g)$$

Now lets pretend that we have a 22.4 liter container at 0 °C with 2 moles of C, 1 mole of B, and 4 moles of A. Assume that this mixture is at equilibrium. We solve for the partial pressure equilibrium constant and get:

$$K = \frac{P_C^{\,2}}{P_A \times P_B^{\,3}} = \frac{2^2}{4 \times 1^3} = 1$$

If you don't understand how we arrived at the partial pressures, go back to the section of this Lecture where we discussed standard molar volume. This is important.

Now we double the pressure with two different methods. First we reduce the volume to 11.2 liters. Notice that the partial pressures have doubled. We can solve for Q before we allow the gasses to react.

$$Q = \frac{P_C^{\,2}}{P_A \times P_B^{\,3}} = \frac{4^2}{8 \times 2^3} = .25$$

Notice that Q is smaller than K. Since Q always moves toward K, Q will get larger. Since Q is products over reactants, products will increase and reactants will decrease. This is the rightward shift predicted by Le Chatelier's principle.

The second method we use to double the pressure is to add 7 moles of D, a nonreactive gas. Notice that, although the total pressure changes, the partial pressures remain the same. Therefore, Q equals K, and there is no shift. Le Chatelier's principle has failed us. This is because Le Chatelier's principle is based on equilibrium.

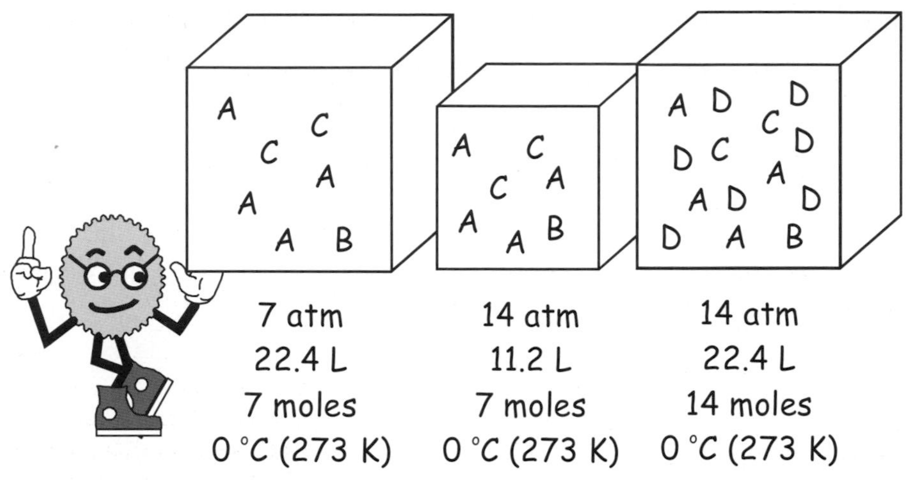

On the MCAT, Le Chatelier's principle works most of the time. However, when using it, you must be careful.

Questions 26 through 30 are **NOT** based on a descriptive passage.

26. As temperature is increased, the equilibrium of a gaseous reaction will always:

A. shift to the right.
B. shift to the left.
C. remain constant.
D. The answer cannot be determined from the information given.

27. All of the following are true concerning a reaction at equilibrium EXCEPT:

A. The rate of the forward reaction equals the rate of the reverse reaction.
B. There is no change in the concentrations of both the products and the reactants.
C. The activation energy has reached zero.
D. The Gibbs free energy has reached a minimum.

28. Nitric acid is produced commercially by oxidation in the Oswald process. The first step of this process is shown below.

$$4NH_3(g) + 5O_2(g) \rightarrow 4NO(g) + 6H_2O(g)$$

A container holds 4 moles of gaseous ammonia, 5 moles of gaseous oxygen, 4 moles of gaseous nitric oxide, and 6 moles of water vapor at equilibrium. Which of the following would be true if the container were allowed to expand at constant temperature?

A. Initially during the expansion the forward reaction rate would be greater than the reverse reaction rate.
B. The equilibrium would shift to the left.
C. The partial pressure of oxygen would increase.
D. The pressure inside the container would increase.

29. Nitrous oxide is prepared by the thermal decomposition of ammonium nitrate.

$$NH_4NO_3(s) \rightarrow N_2O(g) + 2H_2O(g)$$

The equilibrium constant for this reaction is:

A. $[NH_4NO_3]/[N_2O][H_2O]^2$
B. $[N_2O][H_2O]^2/[NH_4NO_3]$
C. $[N_2O][H_2O]^2$
D. $[N_2O][H_2O]$

30. Which of the following is true concerning a reaction that begins with only reactants and moves to equilibrium?

A. The rate of the forward and reverse reactions decreases until equilibrium is reached.
B. The rate of the forward and reverse reactions increases until equilibrium is reached.
C. The rate of the forward reaction decreases, and the rate of the reverse reaction increases until equilibrium is reached.
D. The rate of the forward reaction increases, and the rate of the reverse reaction decreases until equilibrium is reached.

Answers to Questions 26-30

26. D is correct. Equilibrium will probably shift with temperature. The direction is dictated by thermodynamics. We need more information.

27. C is correct. The activation energy is dictated by the reaction itself and doesn't change during the reaction. We will see later that the Gibbs free energy is at a minimum when a reaction is at equilibrium.

28. A is correct. By Le Chatelier's principle the equilibrium would shift to the right causing an increase in the forward reaction.

29. C is correct. The equilibrium constant is products over reactants with the coefficients as exponents. However, reactants and products in different phases have an exponent of zero, so they are not included.

30. C is correct. Initially there are no products, so the reverse reaction begins at zero. As the reactants are used up, the forward reaction slows down. Equilibrium is the point where the rates equalize.

Lecture 3: Thermodynamics

Thermodynamics is the study of energy and its relationship to macroscopic properties of chemical systems. Thermodynamic functions are based on probabilities, and are only valid for systems composed of a large number of molecules. In other words, with few exceptions, the rules of thermodynamics govern complex systems containing many parts, and they cannot be applied to specific microscopic phenomena.

Thermodynamic problems divide the universe into a **system** and its **surroundings**. The system is the macroscopic body under study, and the surroundings are everything else. There are three systems: *open*, *closed*, and *isolated*. System definitions are based upon mass and energy exchange with the surroundings. Open systems exchange both mass and energy with their surroundings; closed systems exchange energy but not mass; and isolated systems do not exchange energy or mass.

Heat

There are only two ways to transfer energy between systems: **heat q** and **work w**. Heat is the transfer of thermal energy (explained below) from a system with a higher temperature to a system with a lower temperature. In other words, heat is the natural movement of energy from a hot body to a cooler body. Any energy transfer that is not heat is work.

Heat has three forms: conduction, convection, and radiation. **Conduction** is thermal energy transfer via molecular collisions. Conduction requires direct physical contact. When one system touches another, and the higher energy molecules of one system transfer some of their energy to the lower energy molecules of the other system through molecular collisions, this is energy transfer called conduction. An object's ability to conduct heat in this fashion is called its thermal conductivity. A slab of a given substance with face area A, length L, and thermal conductivity k will conduct heat Q from a hot body at temperature T_h to a cold body at temperature T_c in an amount of time t.

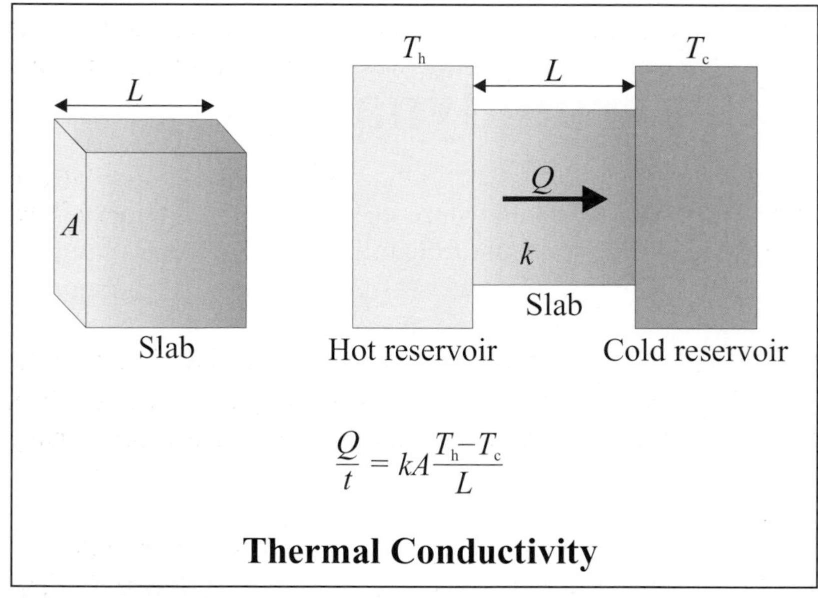

$$\frac{Q}{t} = kA\frac{T_h - T_c}{L}$$

Thermal Conductivity

Q/t is the rate of heat flow. Similar to the rate of fluid flow in an ideal fluid (see Physics Lecture 6) or to the flow of electric current through resisters in series (see Physics Lecture 7), in a *steady state system* the rate of heat flow is constant across any number of slabs between two heat reservoirs. In other words, if the slab in the diagram to the left were replaced by two new slabs with different lengths and thermal conductivities, placed side by side, the rate of heat flow, Q/t, would be the same in both new slabs. This is because energy is conserved. If the rate of energy transfer were not steady across the slabs, there would have to be an energy buildup at the point where the rates were different. This means that the order in which we place the slabs does not affect the overall conductivity. It also means that a higher conductivity results in a lower temperature difference across a slab.

Let see how well you understand thermal conductivity. If I have a heavy blanket and a light blanket, which blanket should I place on top in order to stay warmer? The answer is: the order of the blankets will not make any difference.

If you understand fluids or electricity, you can use them as an analogy to thermal conductivity. The temperature difference is analogous to the pressure difference or the voltage. The rate of heat flow is analogous to volume flow rate or current. Compare Ohm's law for voltage to the analogous equations in thermal conductivity and fluid flow in the box on the right. This should give you some idea of what electrical resistance and fluid resistance depend upon: length and cross-sectional area of the wire or pipe. It should remind you that voltage <u>is</u> a difference and always compares two points, like difference in temperature and difference in pressure.

$$V = i \quad R$$

$$\Delta T = \frac{Q}{t} \; \frac{L}{kA}$$

$$\Delta P = Q \quad R$$

<u>Convection</u> is thermal energy transfer via fluid movements. In convection, a fluid mass moves from one place to another, and carries its thermal energy with it. Ocean and air currents are examples of convection.

<u>Radiation</u> is thermal energy transfer via electromagnetic radiation. Radiation is the only type of heat that transfers through a vacuum. All objects radiate heat. The rate at which an object radiates electromagnetic radiation (the power P) is given by the equation:

$$P = \sigma \varepsilon A T^4$$

where A is the surface area of the object, T is the temperature of the object in kelvins, σ is the Stefan-Boltzman constant (5.67×10^{-8} W m^{-2} K^{-4}), and ε is the *emissivity* of the objects surface, which has a value between 0 and 1. An object with an emissivity of 1 is called a *black body radiator* and is possible only in theory.

In the above equation, if we substitute the temperature of the environment for the temperature of the object, we find the rate at which the object absorbs radiant heat from its environment. The net heat transfer will always be from hot to cold.

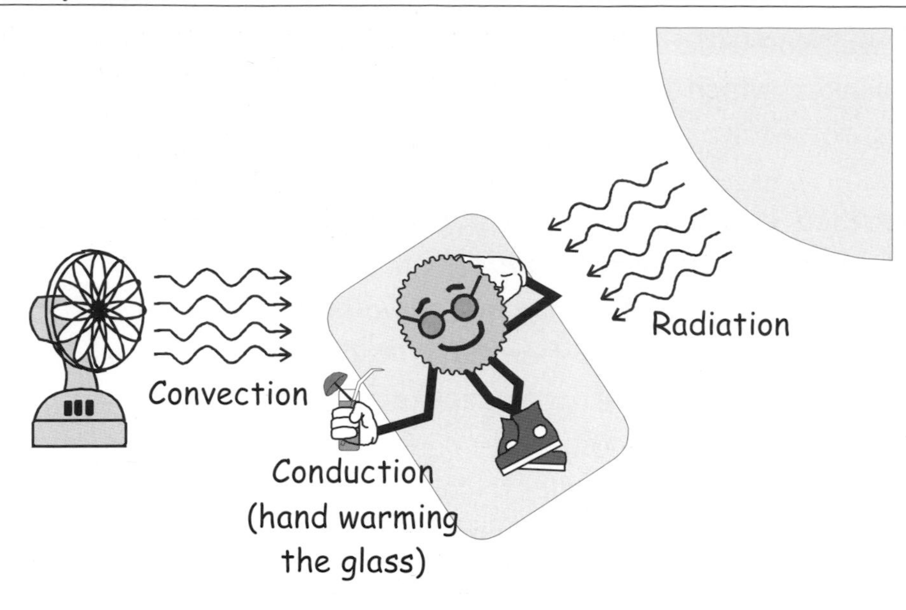

3 types of Heat

A **state function** is a property of the state of a system. Changes in a state function are independent of the path used to achieve them. Notice that heat is not a state function. It does not describe a property of a system, and it does depend upon the pathway of the reaction.

Work

We have already looked at work in Physics Lecture 3, and, from the discussion above, we know that work is any energy transfer that is not heat. However, now we want to look at work as it applies to a chemical system. A system at rest, with no gravitational potential energy and no kinetic energy, may still be able to do ***PV* work**. Imagine a cylinder full of gas compressed by a piston. If we place a mass m on the top of the piston and allow the gas pressure to lift the mass at a constant velocity to a height h, the system has done work on the mass in the amount of the gravitational energy change mgh. Thus negative work has been done on the system. This work is called PV work because, at constant pressure, it is equal to the product of the pressure and the change in volume ($P\Delta V$).

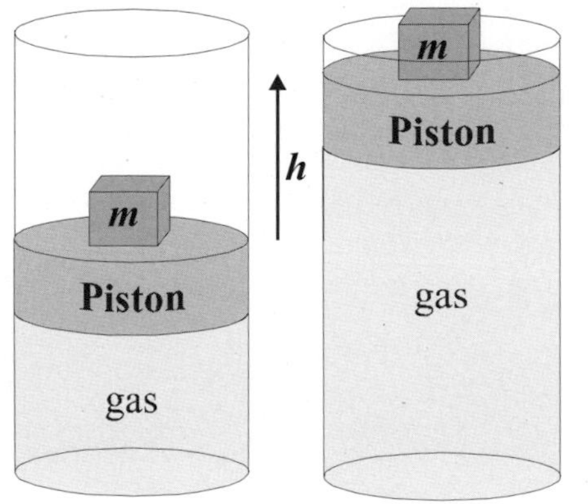

Since the mass is lifted at constant velocity, we know the exact pathway followed from the beginning to the end of the expansion. We know that a constant force of mg must have been applied and therefore the pressure had to remain constant ($F = PA$). According to $PV = nRT$, in order to maintain constant pressure, we must constantly increase the temperature as the volume increases; we must heat the gas.

An MCAT problem involving PV work will probably deal only with a constant pressure situation. However, you should understand that work is not a state function; it is path dependent. If we examine an expanding gas where both pressure and volume are changing, there are an infinite number of paths (combinations of P and V) that we could follow between any two states, and each path may result in a different amount of work. If we were to graph three of these expansions on a P vs. V graph, the area under the curve would give the amount of work done by the system. Notice that, although the beginning states and ending states are the same for each expansion, the area under the curve, and thus the work, is different for all three paths.

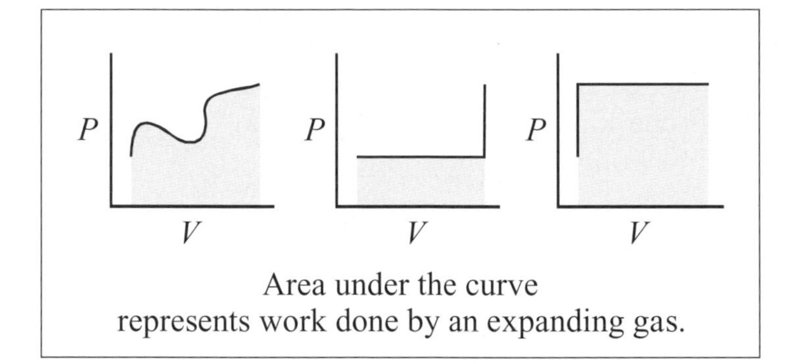

Area under the curve represents work done by an expanding gas.

For the MCAT, you must understand what a state function is. As far as PV work, you will only need to be able to find the work done under constant pressure conditions ($w = P\Delta V$). Realize that no work is done if the volume doesn't change.

<u>**The First Law of Thermodynamics**</u> states that energy is always conserved. Thus any energy change to a system must be the sum of the work done on the system plus the heat flow into the system.

$$\Delta E = q + w$$

Warning: We have chosen the convention where work <u>on</u> the system is positive. It is possible that an MCAT passage could define work done <u>on</u> the system as negative, in which case the formula would be $\Delta E = q - w$.

Notice that if we turn our gas cylinder sideways, and the piston pushes against a mass, we can convert heat to work completely, by maintaining a constant temperature in the gas (assuming ideal conditions). However, to convert heat to work completely on a perpetual basis would require an infinitely long cylinder, and is impossible. If we attempt to bring the cylinder back to its starting point, we must allow heat to flow out of the gas. This is only possible if we can expel the heat to someplace cooler than the gas (a cold reservoir). Since we received the heat from a warmer place (a heat reservoir), we cannot dump the heat back from where it came. (Heat, by definition, is the movement of energy from a warmer body to a cooler body.) This means that in order to rerun the expansion and get more work, some of the energy that entered the gas as heat, must leave as heat. This is one way to state **The Second Law of Thermodynamics**: heat cannot be completely converted into work in a cyclical process.

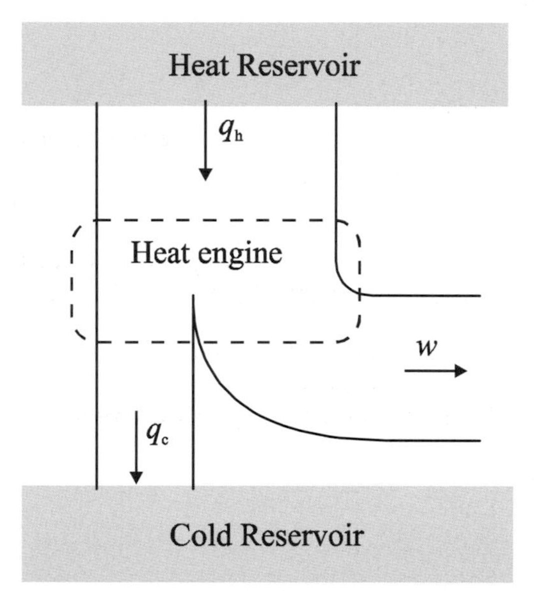

A machine that converts heat to work is called a *heat engine*. A heat engine can be diagrammed as shown. Notice that, via conservation of energy, the heat entering the engine q_h must equal the sum of the work done by the engine w plus the heat leaving the engine q_c.

We can reverse a heat engine to create a refrigerator. If we change the directions of the arrows in the diagram, the inside of the refrigerator is represented by the cold reservoir. Notice that a refrigerator requires work, and that the heat that it generates is greater than the heat that it removes from the cold reservoir.

The most effective cyclical conversion of heat into work is produced by the hypothetical *Carnot engine* and depends upon the temperature difference of the two reservoirs. The further apart the temperatures of the two reservoirs, the more effective the conversion. The fraction of heat that can be converted to work with a Carnot engine is called the efficiency e, and is given by:

$$e = 1 - \frac{T_C}{T_H}$$

where T_H and T_C are the temperatures of the heat and cold reservoirs respectively.

The heat engine stuff is given here in order to help you understand the relationship between heat and work. If it is on the MCAT, it will be explained in a passage. However, don't just ignore it, it is a very possible passage topic and a good way to learn to understand heat and work. At the very least, know the second law of thermodynamics in terms of heat and work: heat cannot be completely converted to work in a cyclical process.

Questions 31 through 35 are **NOT** based on a descriptive passage.

31. Which of the following is true concerning an air conditioner that sits inside a thermally sealed room and draws energy from an outside power source?

A. It requires more energy to cool the room than if part of the air conditioner were outside the room.
B. It will require more time to cool the room than if part of the air conditioner were outside the room.
C. It will require less energy to cool the room than if part of the air conditioner were outside the room.
D. It cannot cool the room on a permanent basis.

32. Three blocks made from the same insulating material are placed between hot and cold reservoirs as shown below.

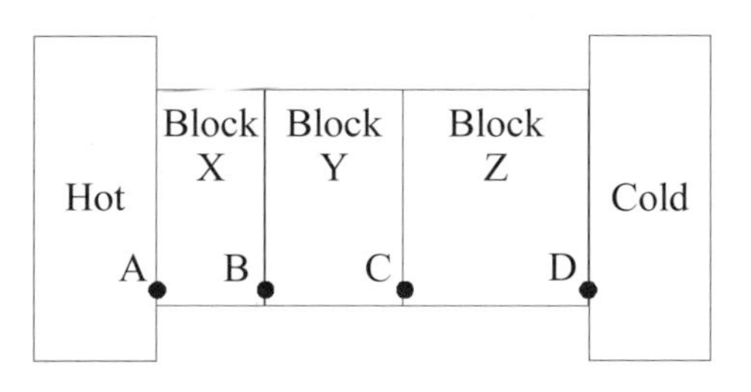

Which of the following must be true?

I. The temperature difference between points A and B is less than the temperature difference between points C and D.
II. The rate of heat flow through Block X is greater than the rate of heat flowing through Block Z.
III. Switching the positions of Block X and Block Z would decrease the rate of heat flow.

A. I only
B. III only
C. I and III only
D. I, II, and III

33. Immediately upon bringing a hot piece of metal into a room, the heat is felt from 5 meters away. The type of heat transfer is probably:

A. convection
B. transduction
C. radiation
D. conduction

34. A box sliding down an incline increases in temperature due to friction. The name for this type of heat is:

A. convection
B. conduction
C. radiation
D. The energy transfer here is due to work and not to heat.

35. Which of the following gas properties is needed to calculate the work done by an expanding gas?

I. The initial and final pressures
II. The initial and final volumes
III. The path followed during the expansion.

A. I and II only
B. I and III only
C. II and III only
D. I, II, and III

Answers to Questions 31-35

31. D is correct. The second law of thermodynamics states that a heat engine cannot have 100% efficiency in converting heat to work in a cyclical process. An air conditioner is a heat engine running backwards. Thus an air conditioner must expel more heat than it takes in when it runs perpetually. A specially made air conditioner could initially cool the room, but to cool the room permanently, it must expel the heat to a heat reservoir.

32. A is correct. I. The temperature difference is inversely proportional to the distance between two points of the same material.

$$\frac{Q}{t} = kA\frac{T_h - T_c}{L}$$

II. The rate of heat flow is constant throughout the blocks, or else heat would build up at the point of slowest flow. III. Since heat flow rate is constant, changing the order of the blocks won't change the rate of heat flow.

33. C is correct. There is no type of heat transfer called transduction. Conduction through the air would take a very long time and be very inefficient. Convection would require some type of air current or breeze. Radiation is as fast as light, and is the correct explanation.

34. D is correct. Unless the box and the incline are at different temperatures, there can be no heat. Energy transfer due to friction is work.

35. D is correct. Work is not a state function, thus we must know the path in order to calculate it.

Thermodynamic Functions

To understand thermodynamics, you must be familiar with seven state functions:

1) internal energy U;
2) temperature T;
3) pressure P;
4) volume V;
5) enthalpy H;
6) entropy S;
7) Gibbs energy G.

Internal Energy

Since thermodynamics is mainly concerned with chemical energy, most problems will not deal with macroscopic mechanical energies. Instead, problems will be concerned with *internal energy*. Internal energy is the collective energy of molecules measured on a macroscopic scale. The energy of molecules includes *vibrational energy, rotational energy, translational energy, electronic energy, intermolecular potential energy, and rest mass energy*. In other words, internal energy is all the possible forms of energy imaginable on a molecular scale. Vibrational energy is created by the atoms vibrating within a molecule. Rotational energy is molecular movement where the spatial orientation of the body changes, while the center of mass remains fixed and each point within a molecule remains fixed relative to all other points. Translational energy is the movement of the center of mass of a molecule. Electronic energy is the potential electrical energy created by the attractions between the electrons and their respective nuclei. The intermolecular potential energy is the energy created by the intermolecular forces between molecular dipoles. Rest mass energy is the energy predicted by Einstein's famous equation $E = mc^2$. The sum of these energies for a group of molecules is called the internal energy.

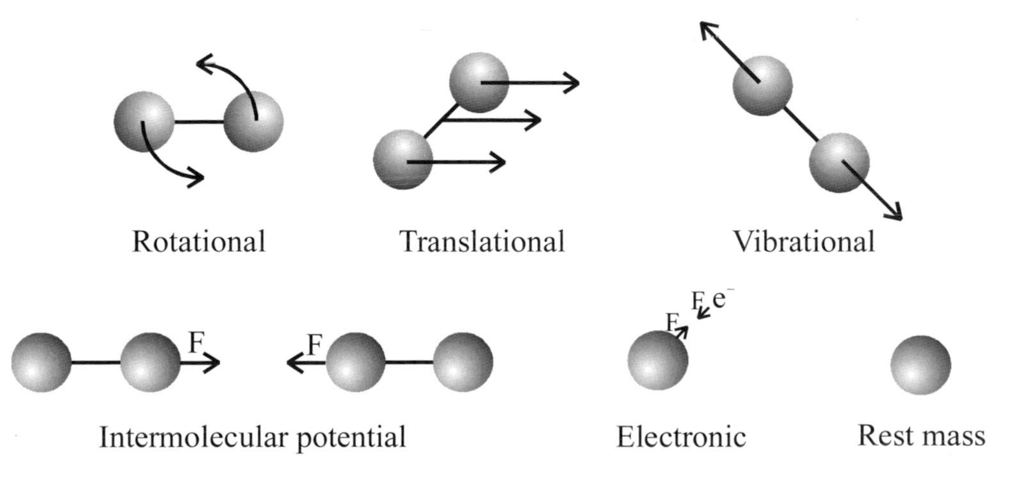

Internal Energy Types

If we have a system at rest, with no electrical or magnetic fields, the only energy change will be in internal energy, and the first law of thermodynamics can be rewritten as: $\Delta U = q + w$. For a reaction within such a system involving no change in volume, the change in internal energy is equal to the heat: $\Delta U = q$ (when the system is closed, and at rest, with no fields).

On the MCAT, you will probably not be tested on internal energy directly. However, in order to understand thermodynamics and conservation of energy, you must understand internal energy. It is a good idea to spend some time thinking about internal energy. The MCAT may refer to internal energy as heat energy or thermal energy. Heat energy and thermal energy are really just the kinetic energy parts of internal energy: vibration, rotation, and translation.

Temperature

The **temperature** of a system is a measure of the average kinetic energy of its molecules. For a fluid, temperature is directly proportional to the translational kinetic energy of its molecules. For solids at high temperatures, temperature is proportional to the kinetic energy of the vibration of its molecules about their equilibrium position. At low temperatures, there is deviation from this rule. (**Warning**: Two systems can have equal internal energies and different temperatures, so temperature is not a direct measure of a system's internal energy.)

For the MCAT, think of temperature as a measure of the random movement of the molecules. For fluids remember the equation: $K.E._{rms} = 3/2\ RT$.

The MCAT will use two measurement systems for temperature: degrees **Celsius** and **Kelvins**. Celsius is just the centigrade system with a new name. At 1 atm, water freezes at 0 °C and boils at 100 °C. The lowest possible temperature is called **absolute zero**, and is approximately –273 °C. To find approximate Kelvins from degrees Celsius, simply add 273. An increase of 1 °C is equivalent to an increase of 1 K.

(Pressure and volume are discussed in Chemistry Lecture 2.)

Enthalpy

When we discussed internal energy above, we assumed no macroscopic mechanical energies to exist, and we included all possible microscopic molecular energies in our definition. However, we did not consider pressure, and pressure has the capacity to perform work. Two systems at rest may have the same amount of internal energy, and, if they are at different pressures, they have different capacities to perform work. (This is one demonstration of why "energy is the capacity to perform work" is a poor definition of energy. See Physics Lecture 3.) **Enthalpy *H*** accounts for both the internal energy of a system and the pressure of that system. Enthalpy of a system is defined as the internal energy of a system and the product of its pressure times its volume.

$$H \equiv U + PV$$

Although enthalpy is given in units of energy (joules), enthalpy itself is not conserved like energy. For instance, in a single reaction, enthalpy can increase for the system and remain unchanged for the surroundings at the same time.

For reactions involving no change in pressure, the change in enthalpy is equal to the heat: $\Delta H = q$ (when the system is closed, at rest, and only PV work is done). Many liquid and solid chemical reactions performed in the lab take place at constant pressure (1 atm) and nearly constant volume. For these reactions (any reactions involving only solids and liquids at moderate pressures) $\Delta U \approx \Delta H$.

Since, in reactions done in the lab, enthalpy often equals heat, the change in enthalpy from reactants to products is often referred to as the **heat of a reaction**.

$$\Delta H^{o}_{\text{reaction}} = \Delta H_{f}^{o}{}_{\text{products}} - \Delta H_{f}^{o}{}_{\text{reactants}}$$

This equation appears to assume an absolute enthalpy value for the products and the reactants. Enthalpy, like most thermodynamic functions, does not have an absolute value associated with it; thus, scientists have assigned enthalpy values based upon **standard states**. Standard state is a somewhat complicated concept that varies with phase (discussed in Chemistry Lecture 4) and other factors. Recall that a 'state' is a specific set of thermodynamic property values. For a pure solid or liquid, the standard state is the state with pressure of 1 bar (about 750 torr or exactly 10^5 pascals) and some temperature T. For a pure gas there is an additional requirement that the gas behave like an ideal gas. An element in its standard state at 25 ^{o}C is arbitrarily assigned an enthalpy value of 0 J/mol. From this value we can assign enthalpy values to compounds based upon the change in enthalpy when they are formed from raw elements in their standard state at 25 ^{o}C. Such enthalpy values for compounds are called **standard enthalpies of formation**. The symbol used to designate standard enthalpy of formation is ΔH_f^{o}. These values can be found by experiment, and are available in books. Standard state values are indicated by the symbol '^{o}' (called naught) following the variable that is to be at standard state.

For the MCAT, standard state will be approximated by 1 atm (760 torr), and the temperature of interest will almost always be 25 °C. For a solution, it will be a 1 molar concentration. So, for the MCAT, you can think of standard state as 1 atm and 25 °C (1 molar concentration for solutions). An element in its standard state at 25 °C is assigned an enthalpy value of 0. For instance H_2 gas and O_2 gas have enthalpy values of 0 because they are elements. 'Enthalpy of formation' is simply the measured enthalpy change when a compound is created from its raw elements at standard state. For instance the enthalpy change for the reaction:

$$H_2 + \tfrac{1}{2}O_2 \rightarrow H_2O$$

is the enthalpy of formation of water.

Since enthalpy is a state function, the change in enthalpy is path independent. Thus, in any reaction, the steps taken to get from reactants to products do not affect the total change in enthalpy. This is called **Hess's law**. Because the enthalpy change is path independent, we can add the enthalpy changes for each step to arrive at the total enthalpy change for a reaction as follows:

$N_2 + O_2 \rightarrow 2NO$	$\Delta H = 180$ kJ	*step 1*
$2NO + O_2 \rightarrow 2NO_2$	$\Delta H = -112$ kJ	*+ step 2*
$N_2 + 2O_2 \rightarrow 2NO_2$	$\Delta H = 68$ kJ	*= complete reaction*

Hess's law also indicates that a forward reaction has exactly the opposite change in enthalpy as the reverse. If the enthalpy change is positive, the reaction is said to be **endothermic**; if it is negative, it is said to be **exothermic**. If we consider a reaction where the change in enthalpy is equal to the heat (a constant pressure reaction), then an exothermic reaction produces heat flow to the surroundings, while an endothermic reaction produces heat flow to the system.

Stated another way "at constant pressure, an exothermic reaction releases heat to the surroundings, and an endothermic reaction absorbs heat from the surroundings". This means that if a reaction is exothermic producing heat flow to the surroundings, the reverse of that reaction is endothermic producing the exact same amount of heat flow to the system.

We can see this graphically if we compare the progress of a reaction with the energy of the molecules. (Due to the close relationship between internal energy and enthalpy, the term energy is used loosely for these types of graphs. You may see the *y* axis labeled as enthalpy, Gibbs free energy, or simply energy.) You can see from the graph below that if the reaction progress is reversed, the enthalpy change is exactly reversed. Notice that there is an initial increase in energy regardless of which direction the reaction moves. This increase in energy is called the **activation energy** (the same activation energy as in Chemistry Lecture 2). The peak of this energy hill represents the molecules in a **transition state** where the old bonds are breaking and new bonds are forming. (In reactions involving more than one molecule, the transition state occurs during the collision.) (Do not confuse the transition state with **intermediates**, which are the products of the first step in a two step reaction). A two step reaction in the diagram similar to the one above would have two humps and two energies of activation. The intermediates would be the trough between the two humps.

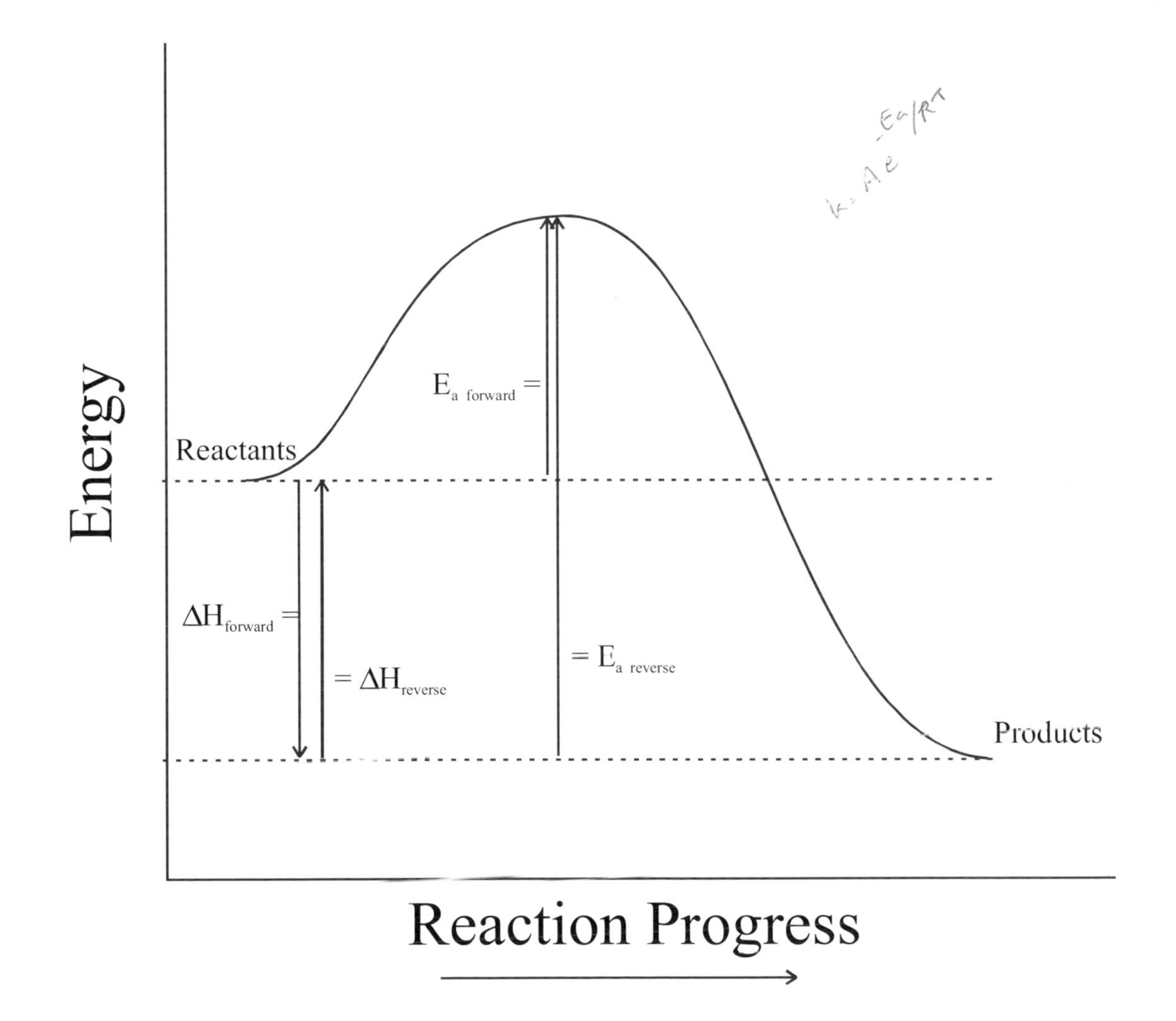

You must understand energy diagrams like the one above.

Notice, on an energy diagram, how a catalyst lowers the activation energy. The activation energy for both the forward and the reverse reactions is lowered. Although the relative amount by which the activation energies are lowered is different, if we used the Arrhenius equation (Chemistry Lecture 2) to find the new rate constants, we would find that the rate constants are raised by the same relative amounts; thus, equilibrium is unaffected by a catalyst. For the MCAT you must remember that a catalyst affects the rate and not the equilibrium. Notice that a catalyst does not affect the enthalpy change either.

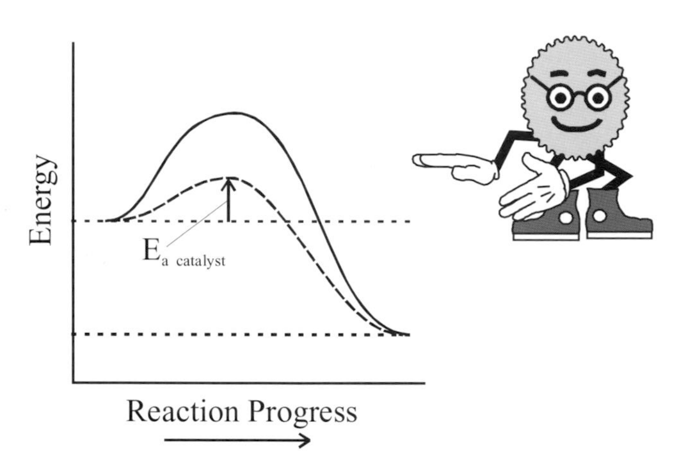

Questions 36 through 40 are **NOT** based on a descriptive passage.

36. What is the enthalpy change in the following reaction?

$$CH_4(g) + 2O_2(g) \rightarrow CO_2(g) + 2H_2O(l)$$

Compound	ΔH_f^0
$CH_4(g)$	–75 kJ/mol
$CO_2(g)$	–394 kJ/mol
$H_2O(l)$	–286 kJ/mol

A. –755 kJ
B. –891 kJ
C. –1041 kJ
D. 891 kJ

37. Which of the following properties of a gaseous system affect its enthalpy?

I. pressure
II. volume
III. internal energy

A. III only
B. I and II only
C. II and III only
D. I, II, and III

38. A catalyst will change all of the following EXCEPT:

A. enthalpy
B. activation energy
C. rate of the forward reaction
D. rate of the reverse reaction

39. In an exothermic reaction, which of the following will most likely increase the ratio of the forward rate to the reverse rate?

A. adding thermal energy to the system
B. removing thermal energy from the system
C. using a catalyst
D. lowering the activation energy

40. The heats of combustion for graphite and diamond are as follows:

$$C_{graphite}(s) + O_2(g) \rightarrow CO_2(g) \quad \Delta H = -394 \text{ kJ}$$
$$C_{diamond}(s) + O_2(g) \rightarrow CO_2(g) \quad \Delta H = -396 \text{ kJ}$$

Diamond spontaneously changes to graphite. What is the change in enthalpy accompanying the conversion of two moles of diamond to graphite?

A. –790 kJ
B. –4 kJ
C. 2 kJ
D. 4 kJ

Answers to questions 36-40

36. B is correct. To find the enthalpy of the reaction we use the following formula:

$$\Delta H^{o}_{reaction} = \Delta H_f^{o}{}_{products} - \Delta H_f^{o}{}_{reactants}$$

The table gives these enthalpies. Don't forget that enthalpy is an extensive process, so quantity matters. We must multiply the enthalpies by the number of moles formed for each molecule. The enthalpy of formation of O_2 is zero, like that of any other molecule in its elemental form at 298 K.

37. D is correct. The definition of enthalpy is: $H \equiv U + PV$

38. A is correct. A catalyst affects the kinetics of a reaction and not the thermodynamics.

39. B is correct. Altering the ratio of the rates of a reaction will change the equilibrium. Removing internal energy from an exothermic reaction will probably push it forward according to Le Chatelier's principle, since heat is a product. Answer C and D concern catalysts and will not change the ratio of the forward and reverse reaction.

40. B is correct. This is Hess's law. We reverse the equation for graphite, so that graphite is a product. In doing so, we must also reverse the sign of the enthalpy. Now we add the two equations and their enthalpies. Don't forget that we must multiply by two for the two moles. Enthalpy is an extensive property.

Entropy

If you have studied **entropy *S*** before, you have probably heard the following example: "Over time, a clean room will tend to get dirty. This is entropy at work." Entropy is nature's tendency toward disorder. A better definition of entropy incorporates the concept of probability. Entropy is nature's tendency to create the most probable situation that can occur within a system. For instance, imagine four identical jumping beans that bounce randomly back and forth between two containers. At any given moment, how many jumping beans are likely to be in the container on the left? If we label each bean A, B, C, and D respectively, we will find that the most likely situation is to have two beans in the container on the left, and one of the least likely situations is to have all four beans in the container on the left. This is because there is only one way for all four beans to be in the container on the left, but there are 6 possible ways that two beans can be in the container on the left. Since any single situation is equally probable, and there are six situations with the two-bean container, the two-bean container is 6 times more probable. Since the greatest possibility exists for the two-bean container situation, this situation has the greatest possible entropy. This type of entropy is called positional entropy. Chemical entropy is actually more complicated than positional entropy; however, for the MCAT, an understanding of positional entropy is sufficient.

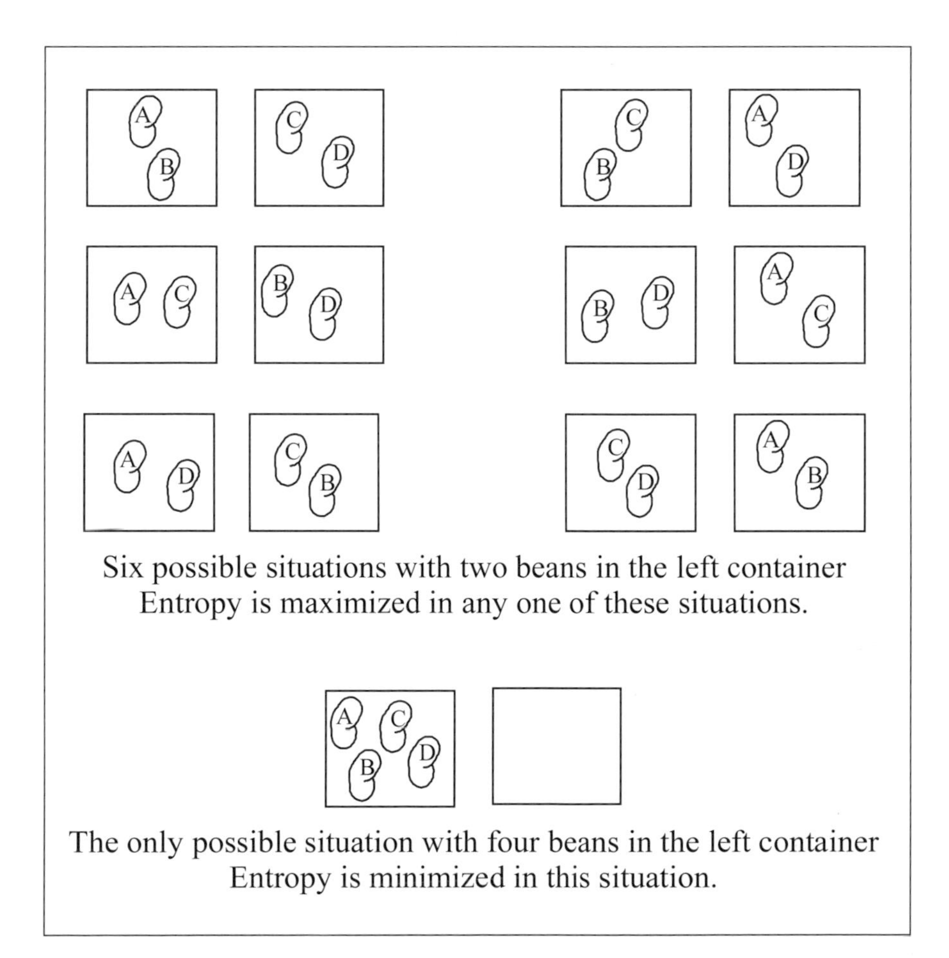

Six possible situations with two beans in the left container
Entropy is maximized in any one of these situations.

The only possible situation with four beans in the left container
Entropy is minimized in this situation.

If we replace the four jumping beans with millions of molecules moving randomly back and forth between the containers, you should be able to see how the odds of having all the molecules on one side become astronomical. The odds are so high, in fact, that **The Second Law of Thermodynamics** states that the entropy of an isolated system will never decrease. We can apply this to any type of system, if we recall that the surroundings of any system include everything that is not in the system. Thus, the system and the surroundings together make up the entire universe. The universe itself is an isolated system. Therefore, the sum of the entropy changes of any system and its surroundings equals the entropy change of the universe, which must be equal to or greater than zero.

$$\Delta S_{system} + \Delta S_{surroundings} = \Delta S_{universe} \geq 0$$

So the entropy of a system can decrease, only if, at the same time, the entropy of the surroundings increases by a greater magnitude. From this equation, we see that a reaction is **reversible** only if its change in entropy for the universe is zero; otherwise, it is **irreversible**. An irreversible reaction must remain infinitely close to a state of equilibrium at all times. (Such conditions are called *quasistatic*.) On a macroscopic scale, all reactions are irreversible. (**Warning:** all chemical reactions are reversible on a molecular scale. Remember, thermodynamics is only concerned with reactions on a macroscopic scale, and, because entropy is defined by probability, entropy is only meaningful on a macroscopic scale. **2nd Warning:** the terms *reversible* and *irreversible* have another meaning on a microscopic scale that is based upon the activation energy. If the activation energy is sufficiently high for one direction of a reaction, the probability that it will occur may be sufficiently low for chemists to call the reaction irreversible. The MCAT is unlikely to adhere to any strict definition of the term 'irreversible'.)

Entropy of the universe is the driving force that dictates whether or not a reaction will proceed. A reaction can be unfavorable in terms of enthalpy, or even energy, and still proceed. But a reaction must increase the entropy of the universe in order to proceed. This explains why reactions proceed to equilibrium. Equilibrium is the point in a reaction where both the system and the surroundings have achieved maximum entropy.

Entropy is a state function. Although there is no absolute entropy, a zero entropy value is assigned for all substances at absolute zero, where molecules have very little motion. The units for entropy are J/K. Entropy change is defined by the infinitesimal change in heat dq_{rev} per Kelvin in a reversible, cyclical process: $\Delta S = dq_{rev}/T$. Entropy can be thought of as the spreading of energy among the molecular energies.

Entropy is an *extensive* property. (It increases with amount of substance.) All other factors being equal, entropy increases with temperature, number, and volume. On the MCAT, if a reaction increases the number of gaseous molecules, then that reaction has positive entropy (for the reaction system, not necessarily for the surroundings or the universe).

The only way that the MCAT is likely to test entropy directly is through the equation given in bold above and the one given on the next page which relates entropy to Gibbs energy. However, you should try to understand entropy qualitatively.

Imagine what happens when you slide a box across the floor. The kinetic energy of the box is dissipated into thermal energy of the box and floor via molecular collisions collectively called friction. Of course, energy is conserved; the increase in thermal energy equals the initial kinetic energy. Now imagine the reverse reaction: the molecules of the floor and the box happen to be moving in a coordinated fashion so as to collide and make the stationary box suddenly start moving. The thermal energy of the molecules becomes kinetic energy. Why doesn't this happen in real life? Energy is still conserved in the reverse reaction, so there is no violation of the first law of thermodynamics. The only reason boxes don't spontaneously start sliding across the floor is due to the decrease in entropy which would accompany such a reaction.

Gibbs Free Energy

Recall that equilibrium is achieved by maximizing the entropy of the system and the surroundings. We can restate this relationship in terms of only the system by using the equation: $\Delta S_{surroundings} = dq_{rev}/T$. From the enthalpy section above, we know that at constant pressure the change in enthalpy is equal to the heat; thus, $\Delta S_{surroundings} = \Delta H_{surroundings}/T$. Also, since $\Delta H_{surroundings} = -\Delta H_{system}$ for these conditions, we have $\Delta S_{surroundings} = -\Delta H_{system}/T$. Therefore, in a closed system capable of doing only *PV* work, and at constant temperature and pressure, equilibrium is achieved by maximizing:

$$\Delta S_{universe} = -\Delta H_{system}/T + \Delta S_{system}$$

If we multiply through by $-T$, and substitute ΔG for $-\Delta S_{universe}/T$, we have the important MCAT equation for **Gibbs free energy *G***:

$$\Delta G = \Delta H - T\Delta S$$

All variables in this equation refer to the system and not the surroundings. This equation is good only for constant temperature reactions, and loses some significance if pressure is not held constant. Before we multiplied the $\Delta S_{universe}$ equation by $-T$, we wanted to maximize $\Delta S_{universe}$ in order to achieve equilibrium. Since we multiplied by a negative to arrive at the Gibbs function, we must minimize the ΔG in the Gibbs function in order to achieve equilibrium. Thus equilibrium is achieved when the change in the Gibbs free energy is zero (constant temperature and pressure, *PV* work only, and a reversible process). A reaction where the change in the Gibbs energy is negative indicates an increase in $\Delta S_{universe}$, and such a reaction is said to occur **spontaneously**. This definition of spontaneity is derived from the Gibbs function, and requires constant temperature and pressure, *PV* work only, and a reversible process. The true definition of a spontaneous reaction requires only that $\Delta S_{universe}$ be positive under any conditions. For the MCAT however, a negative ΔG from the Gibbs function is good enough for spontaneity.

Gibbs energy is an extensive property and a state function. It is not conserved in the sense of the conservation of energy law. An isolated system can change its Gibbs energy. The **Gibbs energy** represents the maximum non-*PV* work available from a reaction, hence the name 'free energy'. Contracting muscles, transmitting nerves, and batteries are some examples of things that do only non-*PV* work, making Gibbs energy a useful quantity when analyzing these systems.

You must know the Gibbs function, and, most importantly, that a negative ΔG indicates a spontaneous reaction. Realize that the Gibbs function deals with the change in enthalpy and entropy.

If a reaction produces a positive change in enthalpy and a negative change in entropy, the reaction can never be spontaneous. Conversely, if a reaction produces a negative change in enthalpy and a positive change in entropy, it must be spontaneous. If the signs of both enthalpy and entropy are the same for a reaction, the spontaneity of the reaction will depend upon temperature. A higher temperature will favor the direction favored by entropy. Remember, these changes are changes in the system and not the surroundings.

$$\Delta G = \Delta H - T\Delta S$$

ΔG	ΔH	ΔS
+	+	-
-	-	+
+/-	-	-
+/-	+	+

Questions 41 through 45 are **NOT** based on a descriptive passage.

41. Which of the following describes a reaction that is always spontaneous?

A. increasing enthalpy and increasing entropy
B. decreasing enthalpy and decreasing entropy
C. increasing enthalpy and decreasing entropy
D. decreasing enthalpy and increasing entropy

42. Which of the following statements about entropy is false?

A. The entropy of a system will always increase in a spontaneous reaction.
B. Entropy is a measure of disorder.
C. The entropy change of a forward reaction is exactly opposite to the entropy of the reverse reaction.
D. Entropy increases with temperature.

43. Which of the following is a violation of the law of conservation of energy?

A. Heat can be changed completely to work in cyclical process.
B. A system undergoing a reaction with constant enthalpy experiences a temperature change.
C. After sliding to a stop, a box with initial kinetic energy K has only thermal energy in an amount less than K.
D. A bond is broken and energy is released.

44. All of the following are examples of processes which increase system entropy EXCEPT:

A. the expanding universe
B. aerobic respiration
C. melting ice
D. building a bridge

45. Which of the following statements is most likely true concerning the reaction:

$$2A(g) + B(g) \rightarrow 2C(g) + D(s)$$

A. System entropy is decreasing.
B. System entropy is increasing.
C. The reaction is spontaneous.
D. The reaction is nonspontaneous.

Answers to Questions 41-45

41. **D is correct.** According to the equation $\Delta G = \Delta H + T\Delta S$, to guarantee that a reaction is spontaneous, enthalpy of the system must decrease and entropy of the system must increase.

42. **A is correct.** The entropy of the universe will increase in a spontaneous reaction. The entropy of a system may or may not increase.

43. **D is correct.** Energy is always required to break a bond.

44. **D is correct.** The process of building a bridge is an ordering process.

45. **A is correct.** Since the number of moles of gas is decreasing with the forward reaction, positional entropy is decreasing. This almost always means that overall system entropy is decreasing. Since the MCAT doesn't distinguish between positional entropy and any other kind of entropy, you can always view a reaction with decreasing number of gas particles as decreasing in entropy and vice versa.

Lecture 4: Solutions

Solutions

A **solution** is a homogeneous mixture of two or more compounds in a single phase, such as solid, liquid, or gas. The MCAT will probably test your knowledge of liquid solutions only. However, you should be aware that solutions are possible in other phases as well. Brass is an example of a solid solution of zinc and copper. Generally, in a solution with two compounds, the compound of which there is more is called the **solvent**, and the compound of which there is less is called the **solute**. Sometimes, when neither compound predominates, both compounds are referred to as solvents. The difference is strictly semantic, and whether a compound is a solute or solvent does not change how it behaves in solution.

There are *ideal solutions*, *ideally dilute solutions*, and *nonideal solutions*. Ideal solutions are solutions made from compounds that have similar properties. In other words, the compounds can be interchanged within the solution without changing the spatial arrangement of the molecules or the intermolecular attractions. Benzene in toluene is an example of a nearly ideal solution because both compounds have similar bonding properties and similar size. In an ideally dilute solution, the solute molecules are completely separated by solvent molecules so that they have no interaction with each other. Nonideal solutions violate both of these conditions. On the MCAT, you can assume that you are dealing with an *ideally dilute* solution unless otherwise indicated; however, you should not automatically assume that an MCAT solution is *ideal*.

The concepts of an ideal and an ideally dilute solution are not tested directly on the MCAT. They are mentioned here in order to deepen your understanding of solutions and to help explain some apparent paradoxes which result when they are not considered.

Colloids

Particles larger than molecules may form mixtures with liquids. If gravity does not cause these particles do not settle out of the mixture over time, the mixture is called a *colloidal suspension*, or *colloid*. Colloidal particles are larger than solute particles, and may even be single large molecules such as hemoglobin. Colloidal suspensions will scatter light, a phenomenon known as the *Tyndall effect*. Colloidal particles are too small to be extracted by filtration; however, heating a colloid or adding an electrolyte may cause the particles to *coagulate*. The larger particles produced by coagulation will settle out or can be extracted by filtration. Colloidal suspensions can also be separated by a semipermeable membrane, a process called *dialysis*.

More Solutions

When a solute is mixed with a solvent, it is said to dissolve. The general rule for dissolution is **'like dissolves like'**. This rule refers to the polarity of the solute and solvent. Highly polar molecules are held together by strong intermolecular bonds formed by the attraction between their partially charged ends. Nonpolar molecules are held together by weak intermolecular bonds resulting from instantaneous dipole moments. These forces are called **London dispersion forces**. A polar solute interacts strongly with a polar solvent by tearing the solvent-solvent bonds apart and forming solvent-solute bonds. A nonpolar solute does not have enough charge to interact effectively with a polar solvent, and thus cannot intersperse itself within the solvent. A nonpolar solute can, however, tear apart the weak bonds of a nonpolar solvent. The bonds of a polar solute are too strong to be broken by the weak forces of a nonpolar solvent.

Ionic compounds are dissolved by polar substances. When ionic compounds dissolve, they break apart into their respective cations and anions and are surrounded by the oppositely charged ends of the polar solvent. This process is called **solvation**. Water acts as a good solvent for ionic substances. The water molecules surround the individual ions pointing their positive hydrogens at the anions and their negative oxygens toward the cations. When several water molecules attach to one side of an ionic compound, they are able to overcome the strong ionic bond, and break apart the compound. The molecules then surround the ion. In water this process is called **hydration**. Something that is hydrated is said to be in an **aqueous phase**. The number of water molecules needed to surround an ion varies according to the size and charge of the ion. This number is called the *hydration number*. The hydration number is commonly 4 or 6.

nitrite	NO_2^-
nitrate	NO_3^-
sulfite	SO_3^{2-}
sulfate	SO_4^{2-}
hypochlorite	ClO^-
chlorite	ClO_2^-
chlorate	ClO_3^-
perchlorate	ClO_4^-
carbonate	CO_3^{2-}
bicarbonate	HCO_3^-
phosphate	PO_4^{3-}

For the MCAT you should be aware of common names, formulae, and charges for the ions listed on the right.

When ions form in aqueous solution, the solution is able to conduct electricity. A compound which forms ions in aqueous solution is called an **electrolyte**. Strong electrolytes create solutions which conduct electricity well and contain many ions. Weak electrolytes are compounds which form few ions in solution.

Units of concentration

There are several ways to measure the concentration of a solution, five of which you should know for the MCAT: **molarity (M), molality (m), mole fraction (χ), mass percentage** and **parts per million (ppm)**. Molarity is the moles of the compound divided by the volume of the solution. Molarity generally has units of mol/L. Molality is moles of solute divided by kilograms of solvent. Molality generally has units of mol/kg and is usually used in formulae for colligative properties (described in Lecture 5). The mole fraction is the moles of a compound divided by the total moles of all species in solution. Since it is a ratio, mole fraction has no units. Mass percentage is 100 times the ratio of the mass of the solute to the total mass of the solution. Parts per million is 10^6 times the ratio of the mass of the solute to the total mass of the solution.

$$M = \frac{\textbf{moles of solute}}{\textbf{volume of solution}}$$

$$m = \frac{\textbf{moles of solute}}{\textbf{kilograms of solvent}}$$

$$\chi = \frac{\textbf{moles of solute}}{\textbf{total moles of all solutes and solvent}}$$

$$\textbf{mass \%} = \frac{\textbf{mass of solute}}{\textbf{total mass of solution}} \times 100$$

$$\textbf{ppm} = \frac{\textbf{mass of solute}}{\textbf{total mass of solution}} \times 10^6$$

Remember that solution concentrations are always given in terms of the form of the solute before dissolution. For instance, when 1 mole of NaCl is added to 1 liter of water, it is approximately a 1 molar solution (not a 2 molar solution even though each NaCl dissociates into two ions).

Normality measures the number of *equivalents* per liter of solution. The definition of an equivalent will depend upon the type of reaction taking place in the solution. The only time normality is likely to appear on the MCAT is with an acid-base reaction. In an acid-base reaction an equivalent is defined as the mass of acid or base that can furnish or accept one mole of protons. For instance, a 1 molar solution of H_2SO_4 would be called a 2 normal solution because it can donate 2 protons for each H_2SO_4. (See Lecture 6 for more on acid-base reactions.)

Questions 46 through 50 are **NOT** based on a descriptive passage.

46. What is the approximate molarity of a NaCl solution with a specific gravity of 1.006?

A. 0.05 *M*
B. 0.06 *M*
C. 0.1 *M*
D. 0.2 *M*

47. Which of the following substances is least soluble in water?

A. NH_3
B. NaCl
C. HSO_4^-
D. CCl_4

48. Which of the following solutions is the most concentrated? (Assume 1 L of water has a mass of 1 kg.)

A. 1 *M* NaCl
B. 1 *m* NaCl
C. an aqueous solution with a NaCl mole fraction of 0.01
D. 55 grams of NaCl mixed with one liter of water.

49. The air we breathe is approximately 21% O_2 and 79% N_2. If the partial pressure of nitrogen in air is 600 torr, then all of the following are true EXCEPT:

A. The mole fraction of nitrogen in air is .79.
B. The mass of nitrogen in a 22.4 L sample of air is 22.4 grams at 0 °C.
C. The partial pressure of oxygen is approximately 160 torr.
D. For every 21 grams of oxygen in an air sample, there are 79 grams of nitrogen.

50. A polar solute is poured into a container with a nonpolar solvent. Which of the following statements best explains the reaction?

A. The strong dipoles of the polar molecules separate the weak bonds between the nonpolar molecules.
B. The dipoles of the polar molecules are too weak to break the bonds between the nonpolar molecules.
C. The instantaneous dipoles of the nonpolar molecules are too weak to separate the bonds between the polar molecules.
D. The instantaneous dipoles of the nonpolar molecules separate the bonds between the polar molecules.

Answers to Questions 46-50

46. C is correct. One liter of water weighs 1 kg; one liter of this solution weighs 1.006 kilograms. If we assume that the volume of water changes very little when NaCl is added, then about 0.006 kg, or 6 g, of NaCl are in each liter of solution. The molecular weight of NaCl is 58.6. 6 grams is about 0.1 moles. (By the way, even if the salt increased the volume of 1 liter of solution by 10 cubic centimeters, the molarity would still be slightly greater than 0.099 *M.* So this is a good approximation. Remember, for dilute solutions, the volume of the solute is negligible.)

47. D is correct. Remember that like dissolves like. Water is polar, and will dissolve polar and ionic substances. A, B, C are ions, ionic compounds, or capable of hydrogen bonding. Carbon tetrachloride is a nonpolar molecule.

48. A is correct. For all practical purposes, the first two are the same. However, since the question asks you to compare them, a one molar solution is 1 mole of NaCl in slightly less than a liter of water. This is because the NaCl requires some volume. A one molal solution is one mole in one full liter of water. (This question assumes that a liter of water has a mass of 1 kg. This is true at 1 atm. and approximately 3 ^{o}C. Water at 1 atm. is at its most dense state at a temperature of slightly over 3 ^{o}C.) There are 55.5 moles of water in a liter (grams/molecular weight = moles). 1/100 = 0.01 and 1/50 = 0.02. Thus a solution with a mole fraction of 0.01 is closer to a 0.5 molar solution than a 1 molar solution. The last answer choice is less than one mole of NaCl in one liter of water.

49. D is correct. You should know that 1 atm is equal to 760 torr. Since the partial pressure of nitrogen is 600, the mole fraction of nitrogen is 0.79. This means that the percentages given are by particle and not by mass. D would be true if the percentages were based on mass. If you chose B, you need to go back to Lecture 3 and review standard molar volume.

50. C is correct. No solution is formed so either B or C must be correct. B is not true.

Solution Formation

The formation of a solution is a physical reaction. It involves three steps: 1) the breaking of the intermolecular bonds between solute molecules; 2) the breaking of the intermolecular bonds between solvent molecules; 3) the formation of intermolecular bonds between the solvent and the solute molecules. Energy is required in order to break a bond. Recall from Chemistry Lecture 3 that at constant pressure the enthalpy change of a reaction equals the heat: $\Delta H \approx q$, and that for condensed phases not at high pressure (for instance the formation of most MCAT solutions) enthalpy change approximately equals internal energy change: $\Delta H \approx \Delta U$. For solution chemistry we shall use these approximations. Thus the heat of solution reactions looks like:

$$\Delta H_{sol} = \Delta H_1 + \Delta H_2 + \Delta H_3$$

Since energy is required to break a bond, the first two steps in dissolution are endothermic while the third step is exothermic.

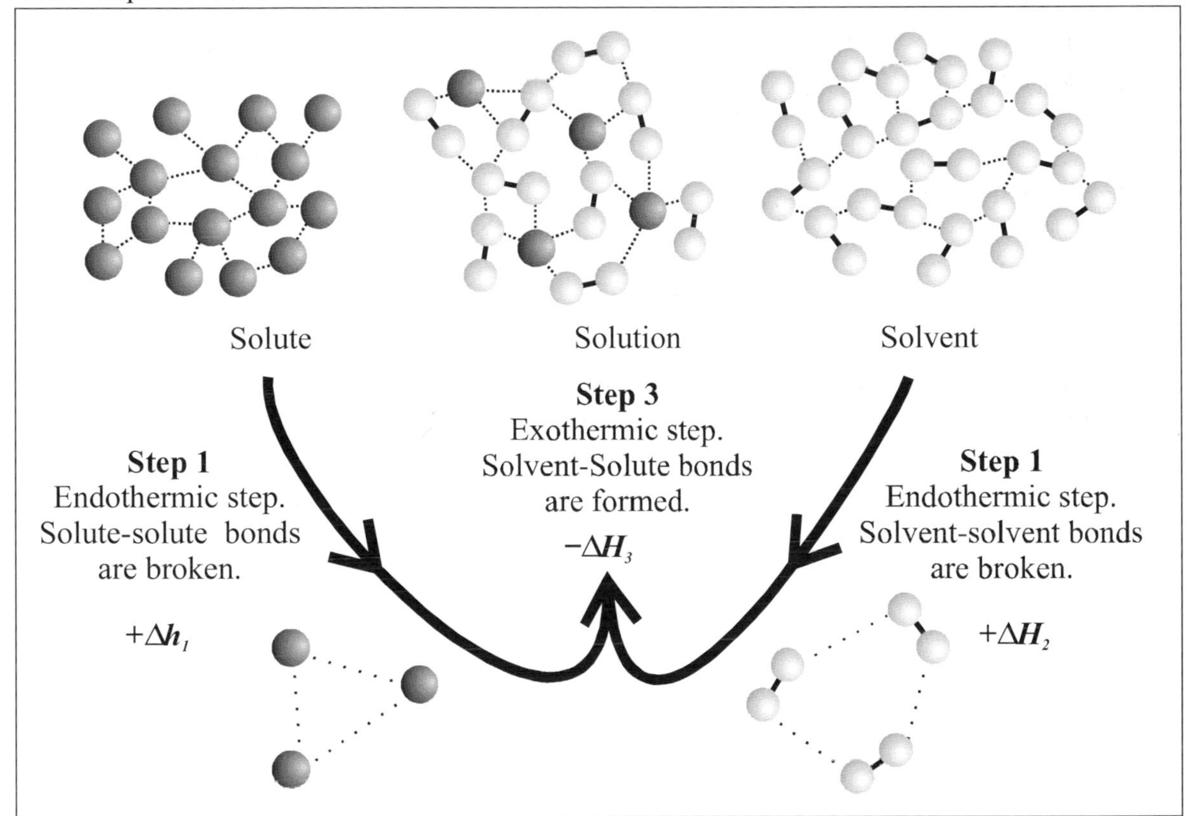

If the overall reaction releases energy (is exothermic), the new intermolecular bonds are more stable than the old, and, in general, the intermolecular attractions within the solution are stronger than the intermolecular attractions within the pure substances. (Remember, less energy in the system usually means a more stable system.) If the overall reaction absorbs energy (is endothermic), the reverse is true. Using the approximations mentioned above, the overall change in energy of the reaction is equal to the change in enthalpy and is called the *heat of solution* ΔH_{sol}. Generally speaking, a negative heat of solution results in stronger intermolecular bonds, while a positive heat of solution results in weaker intermolecular bonds. (Some books combine steps 2 and 3 of solution formation for aqueous solutions calling the sum of their enthalpy changes the *heat of hydration*.)

Since the combined mixture is more disordered than the separated pure substances, most of the time, the formation of a solution involves an increase in entropy. In fact, positional entropy always increases in the formation of a solution, so, on the MCAT, solution formation has positive entropy.

You must recognize that breaking a bond always requires energy input. Since enthalpy and heat are equal at constant pressure, a solution with a negative enthalpy will give off heat when it forms. Thus, a solution that gives off heat when it forms is creating stronger bonds within the solution.

Vapor Pressure

Imagine a pure liquid in a vacuum-sealed container. If we were to examine the space inside the container, above the liquid, we would find that it is not really a vacuum. Instead it would contain vapor molecules from the liquid. The liquid molecules are held in the liquid by intermolecular bonds. However, they contain a certain amount of kinetic energy, which depends upon the temperature. Some of the liquid molecules at the surface contain enough kinetic energy to break the intermolecular bonds that hold them in the liquid. These molecules launch themselves into the open space above the liquid. As the space fills with molecules, some of the molecules crash back into the liquid. When the rate of molecules leaving the liquid equals the rate of molecules entering the liquid, equilibrium has been established. At this point, the pressure created by the molecules in the open space is called the **vapor pressure** of the liquid.

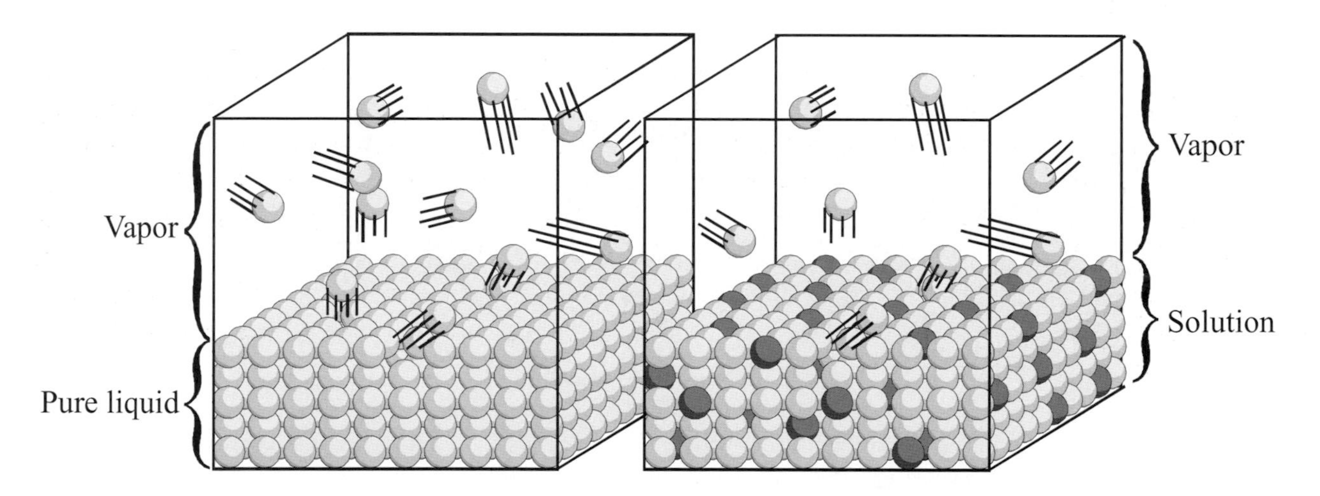

Since vapor pressure is related to the kinetic energy of the molecules, vapor pressure is a function of temperature, and increases with temperature.

When a **nonvolatile solute** (a solute with no vapor pressure) is added to a liquid, some of those solute molecules will reach the surface of the solution, and reduce the amount of surface area available for the liquid molecules. Since the solute molecules don't break free of the solution but they do take up surface area, the number of molecules breaking free from the liquid is decreased while the surface area of the solution and the volume of open space above the solution remain the same. From the ideal gas law, $PV = nRT$, we know that a decrease in n at constant volume and temperature is proportional to a decrease in P. The vapor pressure of the solution P_v is given by **Raoult's law**, and is proportional to the mole fraction of the liquid χ_a and the vapor pressure of the pure liquid P_a.

$$\boldsymbol{P_v = \chi_a P_a}$$

If the solute is a **volatile solute** (a solute with a vapor pressure), the situation is a little more complicated. A volatile solute will also compete for the surface area of a liquid. However, some of the molecules of a volatile solute will escape from solution and contribute to the vapor pressure. If the solution is an ideal solution (solute and solvent have similar properties), the partial pressures contributed by the solvent and solute can be found by applying Raoult's law separately. The sum of the partial pressures gives the total pressure of the solution, and we arrive at a modified form of Raoult's law:

$$\boldsymbol{P_v = \chi_a P_a + \chi_b P_b}$$

where each χP term represents the partial pressure contributed by the respective solvent, and P_v represents the total vapor pressure.

But this is not the entire story. As we saw with heats of solution, if the solution is not ideal, the intermolecular forces between molecules will be changed. Either less energy or more energy will be required for molecules to break the intermolecular bonds and leave the surface of the solution. This means that the vapor pressure of a nonideal solution will deviate from the predictions made by Raoult's law. We can make a general prediction of the direction of the deviation based upon heats of solution. If the heat of solution is negative, stronger bonds are formed, fewer molecules are able to break free from the surface and there will be a negative deviation of the vapor pressure from Raoult's law. The opposite will occur for a positive heat of solution.

The deviation of vapor pressure from Raoult's law can be represented graphically by comparing the mole fractions of solvents with their vapor pressures. Graph 1 below shows only the partial pressure of the solvent as its mole fraction increases. As predicted by Raoult's law, the relationship is linear. Graph 2 shows the vapor pressure of an ideal solution and the individual partial pressures of each solvent. Notice that the partial pressures add at every point to equal the total pressure. This must be true for any solution. Graph 3 and 4 show the deviations of nonideal solutions. The straight lines are the Raoult's law predictions and the curved lines are the actual pressures. Notice that the partial pressures still add at every point to equal the total pressure. Notice also that a positive heat of solution leads to an increase in vapor pressure, and a negative heat of solution, a decrease.

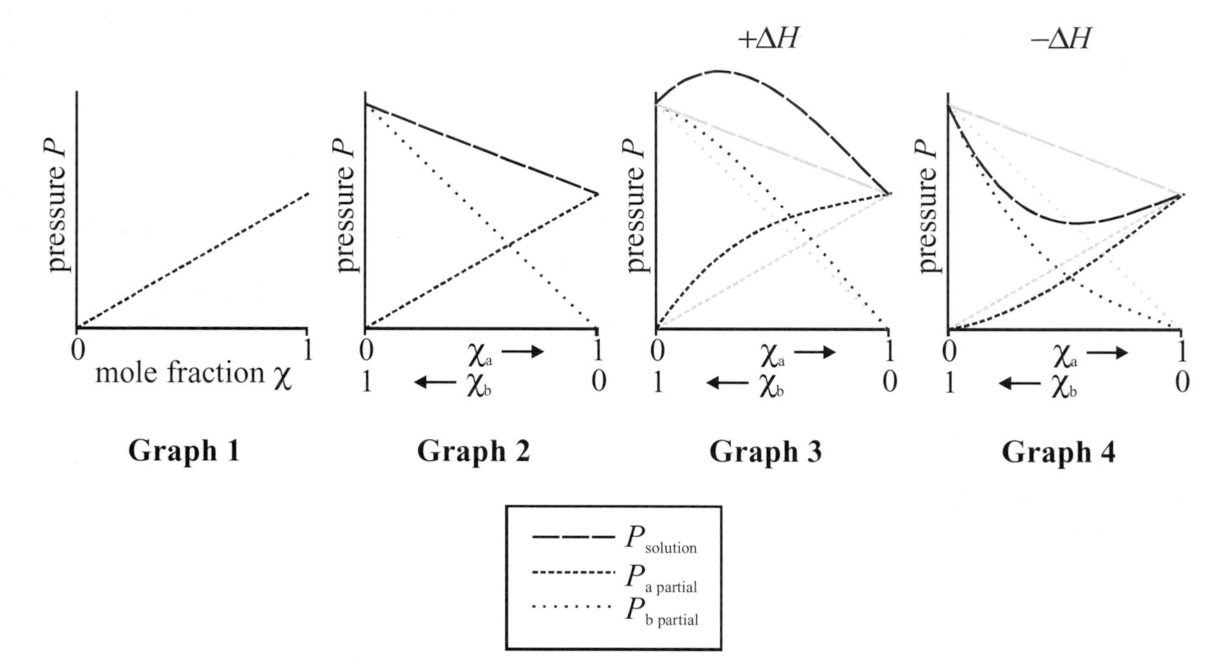

In order to really understand this section you must have a thorough understanding of many of the physics and chemistry concepts that we've studied so far (i.e. bond energy, thermodynamics, pressure, and solutions). I suggest that you reread this section and be sure that you thoroughly understand the concepts.

(For more on vapor pressure see Chemistry Lecture 5)

Questions 51 through 55 are **NOT** based on a descriptive passage.

51. NaCl dissolves spontaneously in water. Based upon the following reaction:

$$NaCl(s) \rightarrow Na^+(g) + Cl^-(g) \quad \Delta H = 786 \text{ kJ/mol}$$

the heat of hydration for NaCl must be:

A. negative with a magnitude less than 786.
B. negative with a magnitude greater than 786.
C. positive with a magnitude greater than 786.
D. Nothing can be determined about the heat of hydration without more information.

52. Which of the following indicates an exothermic heat of solution?

A. Heat is evolved.
B. The final solution is acidic.
C. A precipitate is formed.
D. The reaction is spontaneous.

53. When two pure liquids, A and B, are mixed, the temperature of the solution increases. All of the following must be true EXCEPT:

A. The intermolecular bond strength in at least one of the liquids is less than the intermolecular bond strength between A and B in solution.
B. The reaction is exothermic.
C. The vapor pressure of the solution is less than both the vapor pressure of pure A and pure B.
D. The rms velocity of the molecules increases when the solution is formed.

54. Which of the following will increase the vapor pressure of a liquid?

A. increasing the surface area of the liquid by pouring it into a wider container
B. increasing the kinetic energy of the molecules of the liquid
C. decreasing the temperature of the liquid
D. adding a nonvolatile solute

55. When two volatile solvents are mixed, the vapor pressure drops below the vapor pressure of either solvent in its pure form. What else can be predicted about the solution of these solvents?

A. The solution is ideal.
B. The mole fraction of the more volatile solvent is greater than the mole fraction of the less volatile solvent.
C. The heat of solution is exothermic.
D. The heat of solution is endothermic.

Answers to Questions 51-55

51. D is correct. The change in entropy is positive in solution formation and Gibbs free energy is negative in a spontaneous reaction. From $\Delta G = \Delta H_{sol} - T\Delta S$ we see that the heat of solution may be either positive or negative in this case. The heat of hydration is the separation of water molecules (which requires energy) and the formation of bonds between the ions and water molecules (which releases energy). Thus, the value of the heat of hydration could be either positive or negative. The actual heat of hydration is –783 kJ/mol making the heat of solution +3 kJ/mol.

52. A is correct. At constant pressure, enthalpy is equal to heat.

53. C is correct. The vapor pressure of solution might be lower than just one of the pure substances but not the other. You can see this from the graph below.

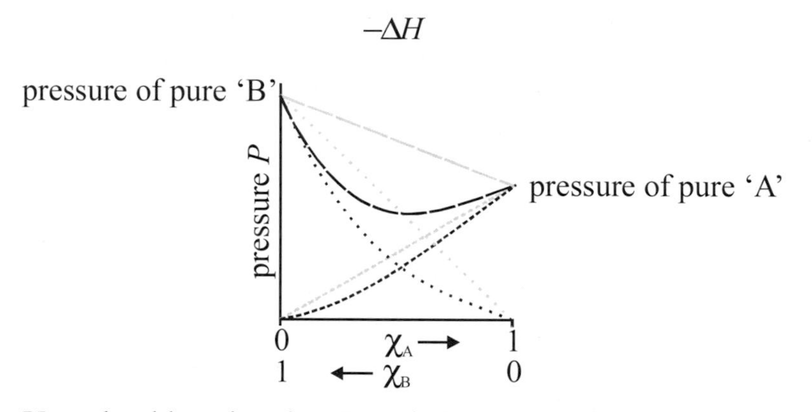

54. B is correct. You should notice that B and C are opposites, so one of them must be the answer. Molecules break free of the surface of a liquid and add to the vapor pressure when they have sufficient kinetic energy to break the intermolecular bonds.

55. C is correct. The solution had to deviate from Raoult's law and therefore could not be ideal. Since it deviated negatively from Raoult's law, the heat of solution is exothermic.

Solubility

Solubility refers to a solute's tendency to dissolve in a solvent. On the MCAT, the solute will usually be a solid. Like any chemical reaction, dissolution of a solid is reversible on a molecular scale. Dissolved molecules of the solid reattach to the surface of the solid. For a dissolving solid, the reverse reaction, called **precipitation**, takes place initially at a slower rate than dissolution. As the solute dissolves and the concentration of solute in solution builds, the rate of dissolution and precipitation equilibrate. At this point, the solution is said to be **saturated**; the concentration of solute has reached a maximum. Just like any other reaction, the equilibrium established at the saturation point is dynamic; the concentrations of products and reactants remain fixed, but the forward and reverse reactions continue at the same rate. The equilibrium of a solution reaction has its own equilibrium constant called the **solubility product K_{sp}**. The solubility product follows exactly the same rules as an ordinary equilibrium constant, and is not different in any way (see Chemistry Lecture 2 for more about equilibrium). Remember that reactants that are not in the same phase as the products receive an exponent of zero in the law of mass action. Thus, solids are left out of the solubility product expression as in the example of the K_{sp} for barium hydroxide shown below.

$$Ba(OH)_2(s) \rightleftharpoons Ba^{2+}(aq) + 2OH^-(aq)$$

$$K_{sp} = [Ba^{2+}][OH^-]^2$$

Solubility and solubility product are not the same thing. The solubility of a substance in a given solvent is found from the solubility product. The solubility is the number of moles per liter of a solute that can be dissolved in a given solvent. Solubility will depend upon the common ions in the solvent. The solubility constant is independent of the common ions in the solvent, and can be found in a reference book.

For most salts, crystallization is exothermic.

We can write an equation for the solvation of BaF_2 in water as follows:

$$BaF_2(s) \rightarrow Ba^{2+}(aq) + 2F^-(aq)$$

The solubility product for BaF_2 is:

$$K_{sp} = [Ba^{2+}][F^-]^2$$

If we look in a book, we find that the K_{sp} for BaF_2 has a value of 2.4×10^{-5} at 25 °C. (The units for any solubility product vary and are found by looking at the equation.) From the K_{sp} we can find the solubility of BaF_2 in any solution at 25 °C. For instance, to find the solubility of BaF_2 in one liter of water, we simply saturate one liter of water with BaF_2. The solubility is the number of moles per liter that dissolve. We call that 'x' for now, since it is unknown. If x moles per liter of BaF_2 dissolve, then there will be x moles per liter of Ba^{2+} in solution and twice as many, or $2x$ moles per liter, of F^-. We plug these values into the K_{sp} equation and solve.

$$2.4 \times 10^{-5} = (x)(2x)^2$$

$$x \approx 1.8 \times 10^{-2}$$

1.8×10^{-2} mol/L is the solubility of BaF_2 in one liter of water at 25 °C.

What would happen if we added 1 mole of F^- to our solution in the form of NaF? The solubility of BaF_2 would change. The NaF would completely dissociate forming 1 mole of F^- and 1 mole of Na^+. The Na^+ ions are not in the equilibrium expression and (ideally) would have no effect on the equilibrium. Because they have no effect, the Na^+ ions are called **spectator ions**. The F^- ions, however, do affect the equilibrium. There disturbance of the equilibrium is called the **common ion effect** because it involves an ion common to an ion in the equilibrium expression. By Le Chatelier's principle, the addition of a common ion will push the equilibrium in a direction which tends to reduce the concentration of that ion. In this case, the equilibrium will move to the left, and the solubility of BaF_2 will be reduced. To find out by exactly how much the solubility will be reduced, we go back to the equilibrium expression. One key to solving solubility problems is realizing that the order in which you mix the solution is irrelevant, so you should mix them in the order that is most convenient to you. In this case it is easiest to add the NaF first, since it completely dissociates. Now we add BaF_2 to a solution of 1 liter of water and 1 mole of F^-. Again, x moles will dissolve leaving x moles of Ba^{2+}. But this time, since there is already 1 mole of F^-, $2x + 1$ moles of F^- will be in solution at equilibrium.

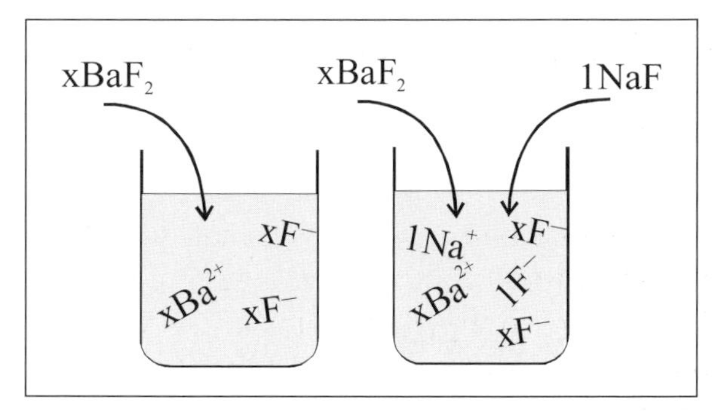

$$2.4\text{x}10^{-5} = (x)(2x + 1)^2$$

Now here's a trick to simplify the math. We know that the equilibrium is shifting to the left, so x will be smaller than before. Even $2x$ will still be much smaller than 1. Thus, $2x + 1$ is going to be very close to 1. Therefore, we drop the $2x$ and solve:

$$2.4\text{x}10^{-5} \approx (x)(1)^2$$

$$x \approx 2.4\text{x}10^{-5}$$

Just to be sure that we can estimate, we plug our estimated value into the term that we deleted ($2x$), and we see if it is truly much smaller than the term to which we added it (in this case 1).

$$2x = 4.8\text{x}10^{-5} << 1$$

Our assumption was valid. Thus our new solubility of BaF_2 is $2.4\text{x}10^{-5}$ mol/L.

Solubility Factors

Pressure and temperature affect solubilities. Pressure on liquids and solids has little effect, but pressure on a gas increases its solubility. For an ideally dilute solution, the increase in pressure is directly proportional to the solubility of a gas, if the gas does not react with, or dissociate in, the solvent. This relationship is given by Henry's law:

$$C = kP$$

where C is the solubility of the gas (typically in moles per liter), k is Henry's law constant, which varies with each solute-solvent pair, and P is the partial pressure of the gas above the solution. The relationship described by Henry's law is easy to remember if we think of canned soda. When we open the can and release the pressure, the solubility of the gas decreases causing some gas to rise out of the solution and create the familiar hiss and foam.

An increase in temperature increases the rate of any reaction, including dissolution. However, it does not necessarily increase the solubility of a solute. Le Chatelier's principle (Chemistry Lecture 2), when applied to solutions, should be used with caution. Although solubilities of solutes in solutions with negative heats may decrease with increasing temperature, because entropy increase is so large in solution formation, the water solubility of most solids increases with increasing temperature. To be absolutely certain, the change in solubility with temperature must be found by experiment. The solubility of gases, on the other hand, typically decreases with increasing temperature. You can remember this by understanding why hot water from factories is hazardous to aquatic life in streams. The hot water has a double effect. First, it holds less oxygen than cold water. Second, it floats on the cold water and seals it off from the oxygen in the air above.

Other factors that affect the solubility of a gas are its size, and reactivity with the solvent. Heavier, larger gases experience greater van der Waals forces and are more soluble. Gases that chemically react with a solvent have greater solubility.

Questions 56 through 60 are **NOT** based on a descriptive passage.

56. When a solution is saturated:

A. the solvent changes to solute, and the solute changes to solvent at an equal rate.
B. the vapor pressure of the solution is equal to atmospheric pressure.
C. the concentration of solvent is at a maximum.
D. the concentration of solvent is at a minimum.

57. The addition of a strong base to a saturated solution of $Ca(OH)_2$ would:

A. decrease the number of OH^- ions in solution.
B. increase the number of Ca^{2+} ions in solution.
C. cause $Ca(OH)_2$ to precipitate.
D. decrease the pH.

58. $NaSO_4$ dissociates completely in water. From the information given in the table below, if $NaSO_4$ were added to a solution containing equal concentrations of aqueous Ca^{2+}, Ag^+, Pb^{2+}, and Ba^{2+} ions, which of the following solids would precipitate first?

Compound	K_{sp}
$CaSO_4$	6.1×10^{-5}
Ag_2SO_4	1.2×10^{-5}
$PbSO_4$	1.3×10^{-8}
$BaSO_4$	1.5×10^{-9}

A. $CaSO_4$
B. Ag_2SO_4
C. $PbSO_4$
D. $BaSO_4$

59. A sealed container holds gaseous oxygen and liquid water. Which of the following would increase the amount of oxygen dissolved in the water?

A. expanding the size of the container
B. adding an inert gas to the container
C. decreasing the temperature of the container.
D. shaking the container

60. The K_{sp} of $BaCO_3$ is 1.6×10^{-9}. How many moles of barium carbonate can be dissolved in 3 liters of water?

A. 4×10^{-5} moles
B. 6.9×10^{-5} moles
C. 1.2×10^{-4} moles
D. 2.1×10^{-4} moles

Answers to Questions 56-60

56. D is correct. Think in terms of mole fraction. The concentration of solvent is at a minimum when the concentration of solute is at a maximum.

57. C is correct. This is the common ion effect (very important for the MCAT).

58. D is correct. We can compare the solubilities in one liter of water. For the compounds that dissociate into two parts, the smallest K_{sp} will be the least soluble and first to precipitate. This is $BaSO_4$. We don't have to compare $BaSO_4$ with Ag_2SO_4 because Ag_2SO_4 dissociates into three particles. This means that if their K_{sp}s were equal, then Ag_2SO_4 would be more soluble than $BaSO_4$. However, the K_{sp} for $BaSO_4$ is much lower, so we know for sure that it is less soluble.

59. C is correct. Gases become more soluble under greater pressure and lower temperatures. The pressure must be the partial pressure of the soluble gas. Adding an inert gas would not change the partial pressure of oxygen in this example. Shaking the can is adding energy, and is similar to heating the can.

60. C is correct. The solubility of $BaCO_3$ in 3 liters of water is found from the equilibrium expression:

$$K_{sp} = [Ba^{2+}][CO_3^{2-}]$$

$$1.6x10^{-9} = [x][x]$$

$$x = 4x10^{-5}$$

This is the saturated concentration in mol/L. We multiply this by 3 liters to get the total number of moles.

Lecture 5: Heat Capacity, Phase Change, and Colligative Properties

Phases

If all the intensive macroscopic properties of a system are constant, that system is said to be *homogeneous*. Any part of a system that is homogeneous is called a *phase*. Some examples of different phases are crystalline solid, amorphous solid, aqueous, pure liquid, and vapor. A system may have a number of solid and liquid phases, but it will usually have only one gaseous phase.

Most of the time, you can think of phases as solid, liquid, and gas. Just be aware that this is not the technical definition. And when we discuss things like solutions, remember that pure water is a different phase then aqueous Na^+ ions. Another common example is rhombic sulfur and monoclinic sulfur; these are two different solid phases of the same element.

Heat Capacity

Phase changes typically arise through changes in internal energy. (This is not to say that the internal energy alone defines the phase of a substance. Equal masses of the same element can have the same internal energy, yet be in different phases.) In the lab, such internal energy changes usually occur through heat. In order to understand phase changes, we must understand how substances react to heat. **Heat capacity C** is a measure of the heat needed to change the temperature of a substance by one degree. Don't let the name 'heat capacity' fool you. Heat is a process, and cannot be stored. Heat capacity was given its name before heat was fully understood. 'Internal energy capacity' would be a better, but not perfect, name.

Since we can heat something without changing the temperature, there must be more parameters to heat capacity than just temperature. For instance, in an isothermal expansion of a gas, we add heat to a gas while expanding it to maintain a constant temperature. Thus, we can heat a gas without changing its temperature. In fact, there are two heat capacities: a **constant volume heat capacity C_V** and a **constant pressure heat capacity C_P**. Remembering the first law of thermodynamics for a system at rest, $\Delta U = q + w$, and remembering that temperature is a function of thermal energy, we can understand one reason why the same substance can have different responses to the same amount of heat. If the volume of a system is held constant, then the system can do no PV work; all energy change must be in the form of heat. This means that none of the heat energy going into the system can escape as work done by the system. Most of the heat energy must go into changing the temperature. On the other hand, when pressure is held constant, some of the heat energy can leave the system as PV work as the volume changes. Thus, at constant pressure, a substance can absorb heat with less change in temperature by expelling some of the energy as work. Therefore, C_P is greater than C_V.

$$C_V = \frac{q}{\Delta T} \quad \text{constant volume}$$

$$C_P = \frac{q}{\Delta T} \quad \text{constant pressure}$$

However, for a solid or liquid, which have very little change in volume, there is a more important reason why the heat capacities differ. The intermolecular forces of a solid or liquid are much stronger than those of a gas. Small changes in the intermolecular distances of noncompressibles result in large changes in intermolecular potential energy. Intermolecular potential energy does not affect temperature, and thus heat is absorbed at constant pressure with less change in temperature than at constant volume. Again: C_P is greater than C_V.

Heat capacity is always positive on the MCAT; the temperature will always increase when heat is added to a substance at constant volume or pressure. In the real world, heat capacity also changes with temperature. However, unless otherwise indicated, for the MCAT, assume that heat capacity does not change with temperature.

Sometimes the MCAT will give you the heat capacity of an entire system. For instance, you may be shown a container of oil with a thermometer and a stirring device. Each of these things has its own heat capacity; the oil, the container, the thermometer, the stirring device, etc. However, for simplicity, the MCAT may give you the heat capacity of the system as a whole. In such a case, the heat capacity will be given in some units of energy divided by some units of temperature: i.e. J/K or cal/°C. For these situations, use the following equation:

$$q = C\Delta T$$

Sometimes the MCAT will give a **specific heat capacity** c. 'Specific' always means divided by mass, so the specific heat capacity is simply the heat capacity per unit mass. A specific heat usually has units of J kg^{-1} K^{-1} or cal g^{-1} $°C^{-1}$. When a specific heat is given, use the following equation:

$$q = mc\Delta T$$

The 'm' in this equation is for mass, not molality. This equation is easy to remember because it looks like q = MCAT. Notice that the symbol for specific heat is a small 'c'. For the MCAT you must know that water has a specific heat of 1 cal g^{-1} $°C^{-1}$. This is the definition of a calorie.

$$c_{water} = 1 \text{ cal g}^{-1} \, °\text{C}^{-1}$$

Just think about the heat capacity of a substance as the amount of heat a substance can absorb per unit of temperature change. Don't worry too much about the difference between constant pressure and constant volume heat capacities. The MCAT might even ignore this fact completely. Use units to help you solve heat capacity problems. For instance, if a heat capacity is given in cal g^{-1} $°C^{-1}$, then you know that to find the heat (measured in calories) you simply multiply by grams and degree Celsius. This gives you the equation $q = mc\Delta T$. Most of the time you don't have to know the formula, if you look at the units. Also with heat capacity problems, follow the energy flow, remembering that energy is always conserved: $\Delta E = q + w$.

By the way, don't be surprised if you see molar heat capacity or something similar. Heat capacities can be given per mole, per volume, per gram, or per whatever. Just use the equation $q = mc\Delta T$ and rely on the units of c to find the units of m. For instance, if c is given as the molar heat capacity, m would be in moles.

Calorimeters

A **calorimeter** is a device which measures heat flow. There are both constant pressure and constant volume calorimeters. A **coffee cup calorimeter** is an example of a constant pressure calorimeter because it measures heat flow at atmospheric pressure. In a coffee cup calorimeter two coffee cups are used to hold the energy inside the solution. A stirrer maintains equal distribution of energy throughout, and a thermometer measures the change in temperature. Obviously, a coffee cup calorimeter cannot contain expanding gases. Reactions that take place inside a coffee cup calorimeter occur at the constant pressure of the local atmosphere. Therefore, it is used to measure the heat of reactions. (Recall that at constant pressure $q = \Delta H$.) For instance, if we mix HCl and NaOH in a coffee cup calorimeter, the net ionic reaction is:

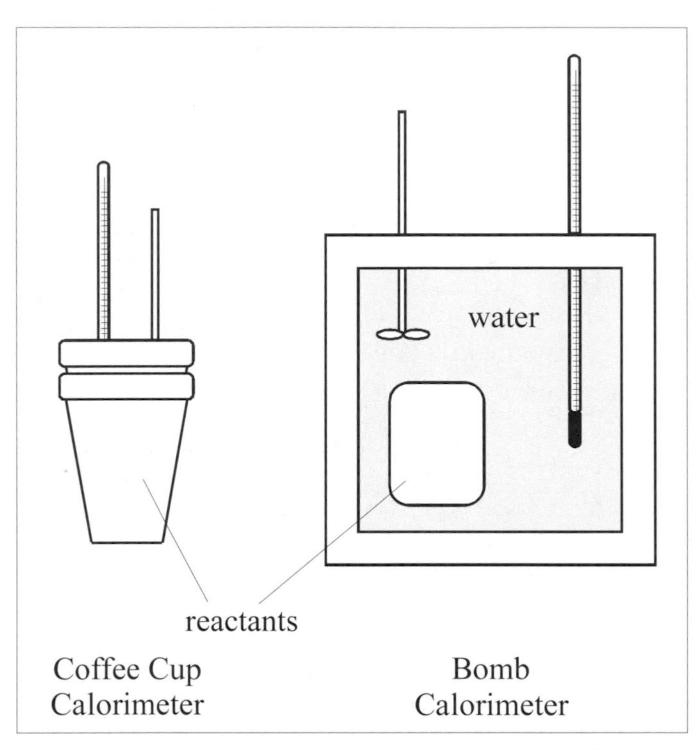

$$H^+ + OH^- \rightarrow H_2O$$

Using the specific heat of water, the mass of water, and the measured change in temperature, we can solve for q in the equation: $q = mc\Delta T$. Since $q = \Delta H$ under these conditions, we have the heat of reaction for the equation above. (Since this is a condensed phase $\Delta H \approx \Delta U$.)

A **bomb calorimeter** measures heat flow at constant volume. A bomb calorimeter tells us the internal energy change in a reaction. (Recall that at constant volume $q = \Delta U$.) In a bomb calorimeter, a steel container full of reactants is placed inside another rigid, thermally insulated container. When the reaction occurs, heat is transferred to the water shown in the diagram. Using the known heat capacity of the container and the equation: $q = C\Delta T$, we can deduce the heat of the reaction, and thus the internal energy change in the reaction.

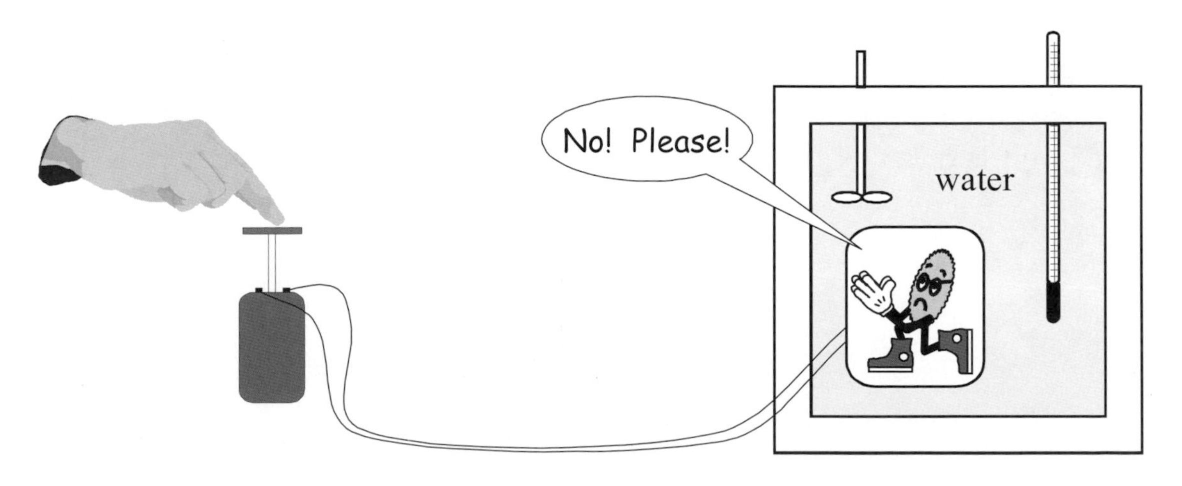

Questions 61 through 65 are **NOT** based on a descriptive passage.

61. 20 grams of NaCl is poured into a coffee cup calorimeter containing 250 ml of water. If the temperature inside the calorimeter drops 1 °C by the time the NaCl is totally dissolved, what is the heat of solution for NaCl and water? (specific heat of water is 4.18 J/g °C.)

A. –3 kJ/mol
B. –1 kJ/mol
C. 1 kJ/mol
D. 3 kJ/mol

62. Using a bomb calorimeter, the change in energy for the combustion of one mole of octane is calculated to be -5.5×10^3 kJ. Which of the following is true concerning this process?

A. Since no work is done, the change in energy is equal to the heat.
B. Since there is no work, the change in energy is equal to the enthalpy.
C. Since work is done, the change in energy is equal to the heat.
D. The work done can be added to the change in energy to find the enthalpy.

63. Which of the following are true statements?

I. The heat capacity of a substance is the amount of heat that substance can hold per unit of temperature.
II. The specific heat for a single substance is the same for all phases of that substance.
III. When heat is added to a fluid, its temperature will change less if it is allowed to expand.

A. I only
B. III only
C. I and III only
D. I, II, and III

64. Substance A has a greater heat capacity than substance B. Which of the following is most likely true concerning substances A and B?

A. Substance A has larger molecules than substance B.
B. Substance B has a lower boiling point than substance A.
C. At the same temperature, the molecules of substance B move faster than those of substance A.
D. Substance A has more methods of absorbing energy than substance B.

65. In a free adiabatic expansion, a real gas is allowed to spread to twice its original volume with no energy transfer from the surroundings. All of the following are true concerning this process EXCEPT:

A. No work is done.
B. Increased potential energy between molecules results in decreased kinetic energy and the gas cools.
C. Entropy increases.
D. The gas loses heat.

Answer to Questions 61-65

61. D is correct. First figure out the heat evolved by the reaction using $q = mc\Delta T$ =>

$$q = 250 \text{ grams x } 4.18 \text{ J/g } ^{\circ}\text{C x } 1\ ^{\circ}\text{C} \approx 1050 \text{ joules}$$

Next divide by moles of NaCl (20 grams is about 1/3 of a mole). This gives you 3150 joules, which is equal to 3 kJ. Since the temperature went down, the reaction is endothermic with positive enthalpy. Notice all the rounding. This problem should have been done with very little math.

62. A is correct. Remember, $\Delta E = w + q$. There is no work done because there is no change in volume in a bomb calorimeter. Thus, the total change in energy is heat. Heat is not enthalpy. Heat equals enthalpy at constant pressure. The pressure is not constant in a bomb calorimeter.

63. B is correct. I is false because objects cannot contain heat, and because the same amount of the same substance can have the same amount of energy and be at different temperatures. Nevertheless, this is a treading the MCAT edge of required knowledge. Don't feel too bad if you chose C. II is false. Different phases will have different specific heats. III is true.

64. D is correct. A, B, and C are false. Temperature is proportional to kinetic energy not just velocity, so more mass per molecule does not make a difference. Boiling point does not make sense; substance A might be water and substance B ice. Answer C mistakenly relies upon speed and not kinetic energy for temperature. D is the correct choice by process of elimination. The more ways that a substance has to absorb energy, the more heat it can absorb with the least change in temperature.

65. D is correct. No energy transfer takes place, so there is no heat or work.

Phase Changes

To understand the phase change process, we shall examine water. If we start with ice at –10 °C and begin heating uniformly and at a constant rate, initially, the energy going into the ice increases the vibration of its molecules and raises its temperature. When the ice reaches 0 °C, the temperature stops increasing while the energy goes into breaking and weakening hydrogen bonds. This results in a phase change; the ice becomes water. When all of the ice has changed phase to water, the temperature begins to rise again. The heat energy goes into increased movement of the molecules. When the water reaches 100 °C, the temperature stops rising and the energy once again goes into breaking hydrogen bonds. This process results in a second phase change: water to steam. Once all the hydrogen bonds are broken, the temperature of the steam begins to rise due to the heat energy, which increases the velocity of the gas molecules. This simplified explanation of phase change and energy is diagrammed below in a **heating curve**. For the MCAT, you must understand the energy changes in this process as explained above.

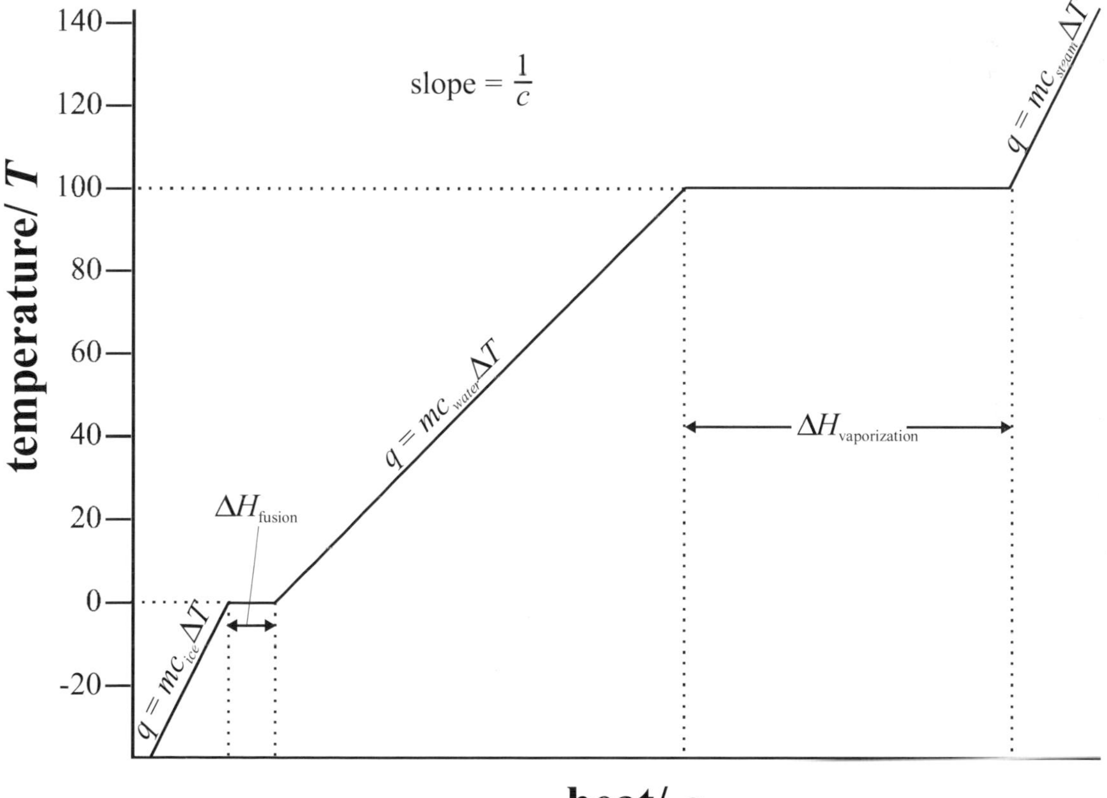

Notice that at 0 °C and 100 °C the heat does not initially change the temperature. At these points, you cannot apply a heat capacity to the substance. This is because at these points, there is a phase change taking place. These points are called the **normal melting point** and **normal boiling point** for water. The word 'normal' indicates a constant pressure of 1 atm. Since the pressure is constant, heat equals the enthalpy change. The enthalpy change associated with melting is called the **heat of fusion**; the enthalpy change associated with boiling, the **heat of vaporization**. Since enthalpy change is a state function, exactly the same amount of heat absorbed during melting is released during freezing. This is also true for vaporization and condensation.

The slope on the heating curve, where not zero, is the reciprocal of the specific heat. Notice that each phase of a substance has a unique slope, and therefore a unique specific heat.

The heating curve shows that melting and boiling are endothermic processes. You should also know that boiling and melting normally increase volume and movement, and therefore result in increased system entropy. From the equation $\Delta G = \Delta H - T\Delta S$, we see that when enthalpy and entropy have the same sign, temperature will dictate in which direction the reaction will move.

Phase Diagrams

A **phase diagram** can be drawn to indicate the phases of a substance at different pressures and temperatures. Each shaded area represents a different phase. The boundaries of the shaded areas represent temperatures and pressures where the corresponding phases are in equilibrium with each other. This equilibrium is also a dynamic equilibrium. For instance, when water and steam are in equilibrium, water molecules are escaping from the liquid phase at the same rate that they are returning. Notice that there is only one point where a substance can exist in equilibrium as a solid, liquid, and gas. This point is called the **triple point**.

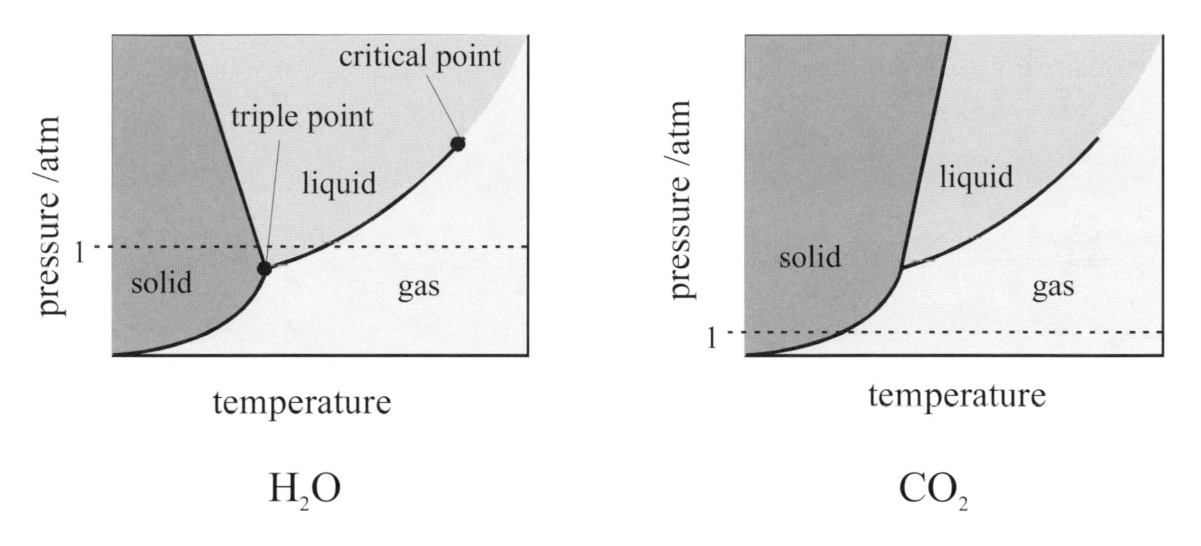

There is also a temperature above which a substance cannot be liquefied regardless of the pressure applied. This temperature is called the **critical temperature**. The pressure required to produce liquefaction while the substance is at the critical temperature is called the **critical pressure**. Together, these two parameters define the **critical point**. Fluid beyond the critical point has characteristics of both gas and liquid, and is called *supercritical fluid.*

Comparing the phase diagrams for water and carbon dioxide, we notice some interesting things. Even if it were not labeled, we could approximate the location of the 1 atm mark for either diagram. We know that at atmospheric pressure, water exists in all three phases at different temperatures. Thus, we know that the 1 atmosphere mark must be above the triple point. Since carbon dioxide (dry ice) sublimes (changes from solid to gas) at one atmosphere, we know that the triple point must be above the 1 atm mark.

Compare the equilibrium line separating the liquid and solid phases on each diagram. For water, the line has a negative slope; for carbon dioxide, a positive slope. Most phase diagrams resemble carbon dioxide in this respect. The negative slope of water explains why ice floats. Since volume decreases with increasing pressure, as we move upward on the phase diagram from ice to liquid water, ice and water must be decreasing in volume and thus increasing in density. Water must be denser than ice. The reason for this is that the crystal structure formed by ice requires more space than the random arrangement of water molecules.

For phase changes you must know where the energy goes. It enters the substance as heat, but what then? During a phase change, it breaks bonds and doesn't change the temperature; otherwise, it increases molecular movement, which increases the temperature.

You must be able to read heating curves and phase diagrams. These two sections are important to understand for a good score on the MCAT.

Think about this: for a single sample of a substance, *P*, *V*, and *T* are interrelated in such a way that if you know two of them, you can derive the other. This means that a phase diagram can also be given in terms of volume. What would that look like? See the problems on the next page for the answer.

Questions 66 through 70 are **NOT** based on a descriptive passage.

66. What is the total heat needed to change 1 gram of water from –10 °C to 110 °C at 1 atm? (ΔH_{fusion} = 80 cal/g, $\Delta H_{vaporization}$ = 540 cal/g, specific heat of ice and steam are 0.5 cal/g °C)

A. –730 cal
B. –630 cal
C. 630 cal
D. 730 cal

67. When heat energy is added evenly throughout a block of ice at 0 °C and 1 atm, all of the following are true EXCEPT:

A. The temperature remains constant until all the ice is melted.
B. The added energy increases the kinetic energy of the molecules.
C. Entropy increases.
D. Hydrogen bonds are broken.

68. Below is a phase diagram for carbon dioxide. What is the critical temperature for carbon dioxide?

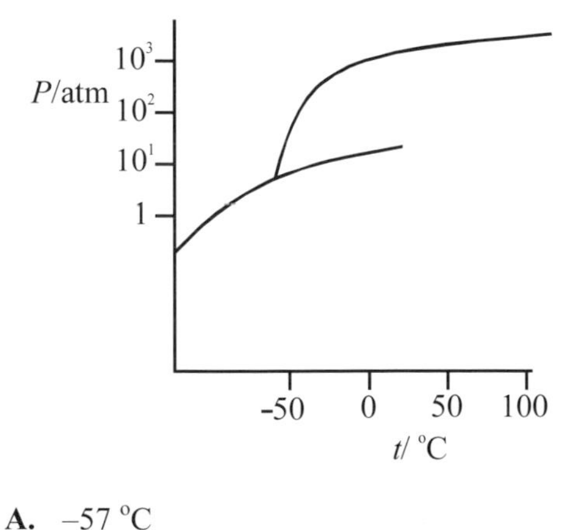

A. –57 °C
B. 0 °C
C. 31 °C
D. 103 °C

69. The diagram below compares the density of water in the liquid phase with its vapor phase.

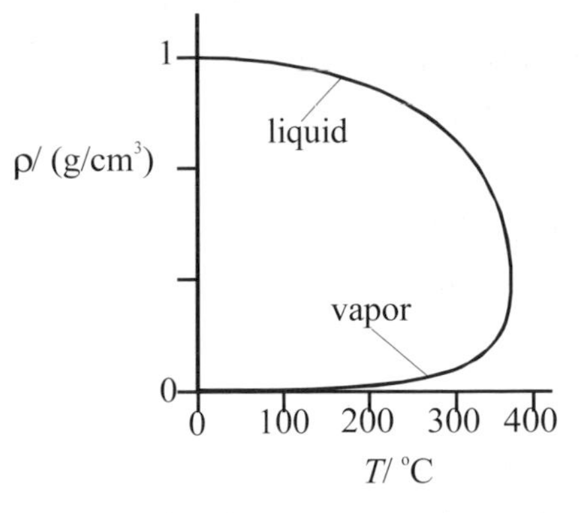

What is the critical temperature of water?

A. 0 °C
B. 135 °C
C. 374 °C
D. 506 °C

70. In graph (a) below, isotherms for water are plotted against pressure and volume. Graph (b) is a phase diagram of water with pressure vs. temperature.

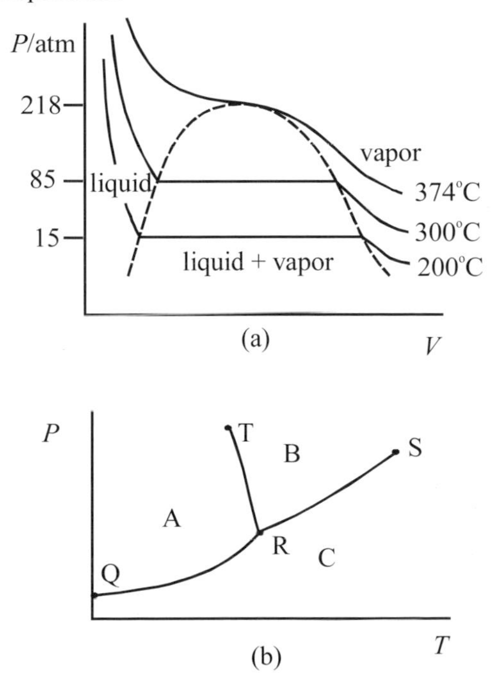

The area inside the dashed line on graph (a) is represented on graph (b) by:

A. the line between points R and S.
B. the area B.
C. the area C.
D. parts of the both area B and C.

Answers to Questions 66-70

66. D is correct. We can solve this problem by summing the q's on the heat curve. The heat is positive because heat is added to the system.

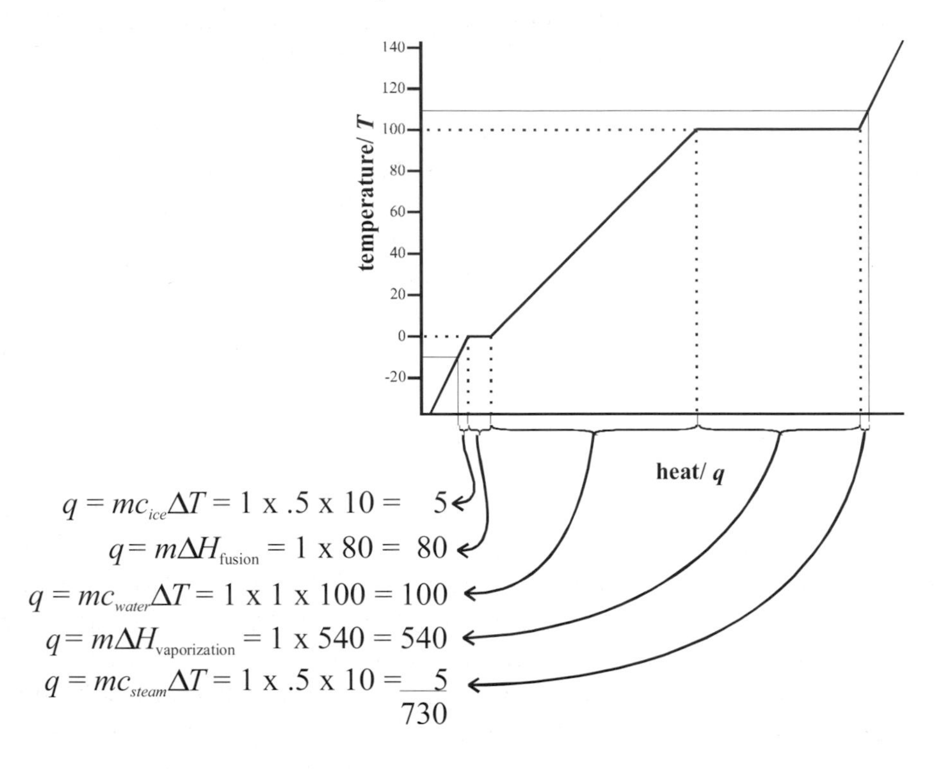

67. B is correct. The added energy goes into breaking bonds, and as is demonstrated by the heat curve above, the temperature remains constant until all the ice is melted. Entropy increases moving to the right on the heat curve.

68. C is correct. This is just a phase diagram with pressure on a log scale. There are many ways to manipulate the phase diagram. Don't be intimidated. Try to compare it to what you know.

69. C is correct. Above the critical point, liquid and vapor water have the same density. The critical temperature will be the highest temperature on the graph where the two lines meet.

70. A is correct. The area in the dashed line is the point where water is changing phase. Like along line RS, in the dashed line area water and steam exist in equilibrium.

Colligative Properties

Some properties in chemistry depend solely on the number of particles, irrespective of the type of particle. Such properties are called **colligative**. There are four colligative properties of solutions: vapor pressure, boiling point, freezing point, and osmotic pressure.

In Chemistry Lecture 4, we saw that the addition of a nonvolatile solute will lower the vapor pressure of the solution in direct proportion to the number of particles added, as per Raoult's law. The vapor pressure has an important relationship to the normal boiling point. When the vapor pressure of a solution reaches the local atmospheric pressure, boiling occurs. Thus, the boiling point of a substance is also changed by the addition of a solute. If that solute is nonvolatile, the vapor pressure is always lowered and the boiling point is always elevated. The equation for the **boiling point elevation** of an ideally dilute solution due to the addition of a nonvolatile solute is:

$$\Delta T = k_b m i$$

where k_b is a specific constant of the substance being boiled, m is the molality of the solution, and i, called the *van't Hoff factor*, is the number of particles into which a single solute particle will dissociate when added to solution. The van't Hoff factor has two possible values: the *expected value* and the *observed value*. The expected value is the value for an ionic compound, if that ionic compound totally dissociates into its respective ions. Thus, the expected value of i for NaCl would be 2; for $MgCl_2$, 3. These values are for an ideally dilute solution. It turns out that, in a nonideal solution consisting of ions, there is *ion pairing*. Ion pairing is the momentary aggregation of two or more ions into a single particle. Ion pairing is not the solute incompletely dissolving; ion pairs are still in the aqueous phase. Ion pairs occur due to the strong attraction between positive and negative ions. The observed value of the van't Hoff factor will take into account ion pairing. Ion pairing increases with solution concentration, and decreases with increasing temperature. In a dilute solution, the observed value will be slightly less than the expected value. On the MCAT, use the expected value unless otherwise instructed.

You cannot apply the equation above to volatile solutes. As shown in Chemistry Lecture 4, a volatile solute can actually increase the boiling point. If you know the heat of solution, you can make qualitative predictions about the boiling point change when a volatile solute is added. For instance, since you know that an endothermic heat of solution indicates weaker bonds, which lead to higher vapor pressure, you can predict that the boiling point will go down.

Melting point also changes when a solute is added, but it is not related to the vapor pressure. Instead, it is a factor of crystallization. Impurities (the solute) interrupt the crystal lattice and lower the freezing point. **Freezing point depression** for an ideally dilute solution is given by the equation:

$$\Delta T = k_f m i$$

Again, the constant k_f is specific for the substance being frozen.

Be careful with freezing point depression. If you add a liquid solute, the impurities will initially lower the melting point; however, as the mole fraction of the solute increases, you will come to a point where the solvent becomes the impurity preventing the solute from freezing. At this point, additional solute acts to reduce the impurities creating a more pure solute, and the freezing point of the solution will rise as solute is added.

The final colligative property is **osmotic pressure**. Osmotic pressure is a measure of the tendency of water (or some other solvent) to move into a solution via osmosis. To demonstrate osmotic pressure, we divide a pure liquid by a membrane that is permeable to the liquid but not to the solute as shown in the diagram. We then add solute to one side. Due to entropy, nature wants to make both sides equally dilute. Since the solute cannot pass through the barrier to equalize the concentrations, the pure liquid begins to move to the solution side. As it does so, the solution level rises and the pressure increases. Eventually a balance between the forces of entropy and pressure is achieved. The extra pressure on the solution side is called osmotic pressure. Osmotic pressure Π is given by:

$$\Pi = iMRT$$

where M is the molarity of the solution.

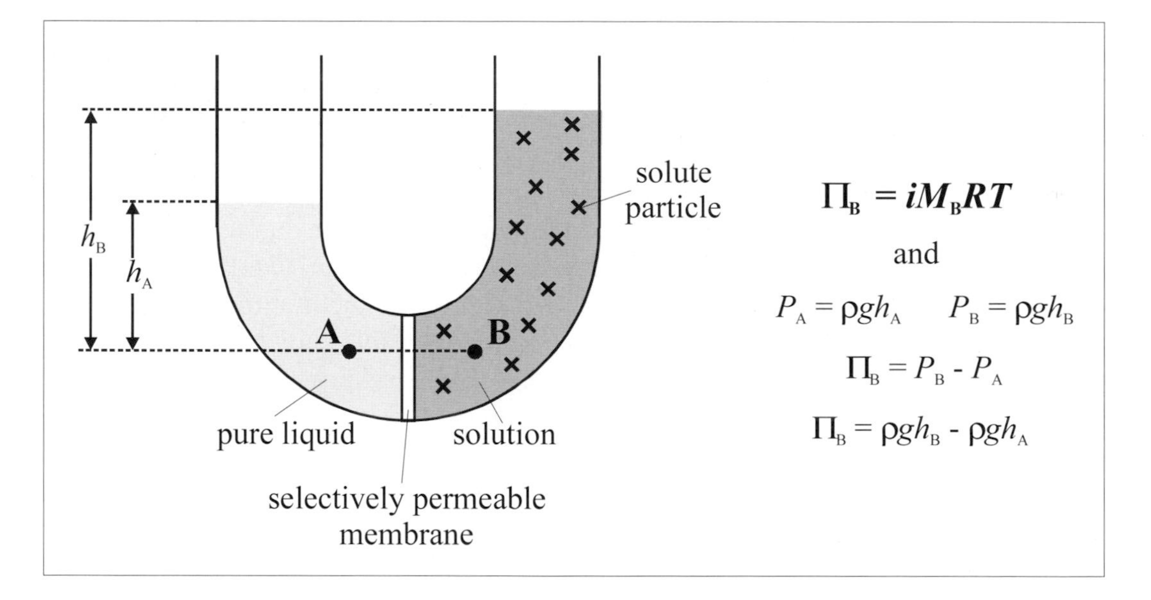

Related to osmotic pressure is *osmotic potential.* Osmotic potential is a partial measure of a system's free energy. Pure water is arbitrarily assigned an osmotic potential of zero. When a solute is added, the osmotic potential becomes negative. At constant temperature and pressure, water flows from higher osmotic potential to lower osmotic potential. *Water potential*, another related term, is similar to osmotic potential but takes into account temperature and pressure. Water potential is essentially the same as free energy. When solutions A and B in the diagram above have come to equilibrium, points A and B have the same water potential, but the osmotic potential of point B is less than that of point A.

MCAT may give the colligative property equations without including the van't Hoff factor, and expect you to know that it needs to be considered.

Questions 71 through 75 are **NOT** based on a descriptive passage.

71. Which of the following aqueous solutions will have the lowest boiling point?

A. 0.5 *M* glucose
B. 1 *M* glucose
C. 0.5 *M* NaCl
D. 0.6 *M* NaCl

72. An object experiences a greater buoyant force in seawater than in fresh water. The most likely reason for this is:

A. Seawater has greater osmotic pressure making the pressure difference greater at different depths.
B. Fresh water has greater osmotic pressure making the pressure difference greater at different depths.
C. Seawater has greater density.
D. Fresh water has greater density.

73. Glycol ($C_2H_6O_2$) is the main component in antifreeze. What mass of glycol must be added to 10 liters of water to prevent freezing down to -18.6 °C? (The molal freezing point depression constant for water is 1.86 °C kg/mol.)

A. 3.1 kg
B. 6.2 kg
C. 10 kg
D. 12.4 kg

74. A student holds a beaker of pure liquid A in one hand and pure liquid B in the other. Liquid A has a higher boiling point than liquid B. When the student pours a small amount of liquid B into liquid A, the temperature of the solution increases. Which of the following statements is true?

A. The boiling point of the solution is lower than either pure liquid A or B.
B. The boiling point of the solution is higher than either pure liquid A or B.
C. The freezing point of the solution is higher than either pure liquid A or B.
D. The vapor pressure of the solution is higher than pure liquid B.

75. 500 ml of an aqueous solution having a mass of 503 grams and containing 20 grams of an unknown protein was placed into a bulb and lowered into pure water as shown. A membrane permeable to water but not to the solute separated the solution from the water.

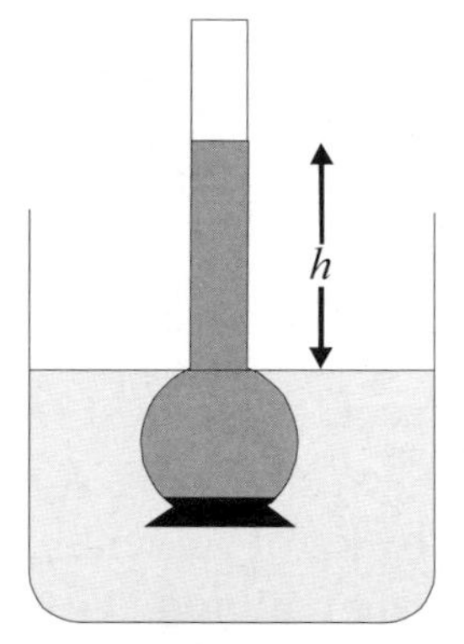

The height of the column of solution was found to be '*h*'. Which of the following statements is true concerning this procedure?

A. A large value for *h* indicates a low osmotic pressure in the solution.
B. A large value for *h* indicates a high osmotic pressure in the pure water.
C. A large value for *h* indicates that the protein has a low molecular weight.
D. A large value for *h* indicates that the protein has a high molecular weight.

Answers to Questions 71-75

71. A is correct. Boiling point elevation is a colligative property. The more particles the higher the boiling point. NaCl dissociates so that the normality is twice the molarity. Thus, the least number of particles will be in 0.5 *M* glucose solution.

72. C is correct. The osmotic pressure will not create a difference in the buoyant force. The equation for buoyant force ($F_b = \rho Vg$) does not include osmotic pressure. Seawater has greater density because salts are heavier than water, and the salt added does not create an appreciable difference in volume.

73. B is correct. You must recognize from the formula that glycerol does not dissociate; it is not ionic. Then use $\Delta T = K_f m$, which gives you a molality of 10. Molality is moles of solute divided by kg of solvent. Assume that 1 liter of water has a mass of 1 kg. Thus 100 moles of glycerol are required. Glycerol has a molecular weight of 62 g/mol. 6200 g = 6.2 kg.

74. B is correct. The reaction is exothermic because the temperature increased. An exothermic reaction makes stronger bonds. Stronger bonds lower vapor pressure. A lower vapor pressure means more energy is needed to raise the vapor pressure to equal atmospheric pressure. Thus, a lower vapor pressure means a higher boiling point.

75. C is correct. The question has a lot of extra information to mislead you. A high value for h indicates a high osmotic pressure in the solution. From the formula for osmotic pressure, $\Pi = iMRT$, we know that a high osmotic pressure corresponds to a high molarity. A high molarity means many particles per gram of protein placed into the solution. Thus a high osmotic pressure means a low molecular weight.

Lecture 6: Acids and Bases

Definitions

There are three definitions of an acid that you must know for the MCAT: Arrhenius, Bronsted-Lowry, and Lewis. These definitions are given here in the order in which they were created. An **Arrhenius acid** is anything that produces hydrogen ions in aqueous solution, and an Arrhenius base is anything that produces hydroxide ions in aqueous solution. This definition covers only aqueous solutions. **Bronsted and Lowry** redefined acids as anything that donates a proton, and bases as anything that accepts a proton. Finally, the **Lewis** definition is the most general, defining an acid as anything that accepts a pair of electrons, and a base as anything that donates a pair of electrons. The Lewis definition includes all the acids and bases in the Bronsted-Lowry and more. Lewis acids include molecules that have an incomplete octet of electrons around the central atom, like $AlCl_3$ and BF_3. They also include all simple cations except the alkali and the heavier alkaline earth metal cations. The smaller the cation and the higher the charge, the stronger the acid strength. Fe^{3+} is a common example of a Lewis acid. Molecules that are acidic only in the Lewis sense are not generally called acids unless they are referred to explicitly as Lewis acids.

Notice that in the Bronsted definition, the acid 'donates', and in the Lewis definition the acid 'accepts'.

Although you must memorize the definitions, it is usually convenient to think of an acid as H^+ and a base as OH^-. In fact, aqueous solutions always contain both H^+ and OH^-. An aqueous solution containing a greater concentration of H^+ than OH^- is acidic, while an aqueous solution containing a greater concentration of OH^- than H^+ is basic. An aqueous solution with equal amounts of H^+ and OH^- is neutral.

H^+	OH^-
acid	base

You should recognize a **hydronium ion H_3O^+**. The hydronium ion is simply a hydrated proton. For MCAT acid and base reactions, a hydronium ion and a proton are the same thing.

Acids taste sour or tart; bases taste bitter. Bases are slippery when wet.

One measure of the hydrogen ion concentration is called the **pH**, where p(x) is a function in which, given any x, p(x) = –log(x). If we measure the hydrogen ion concentration in moles per liter ([H^+] the brackets always indicate concentration), pH is:

$$pH = -\log[H^+]$$

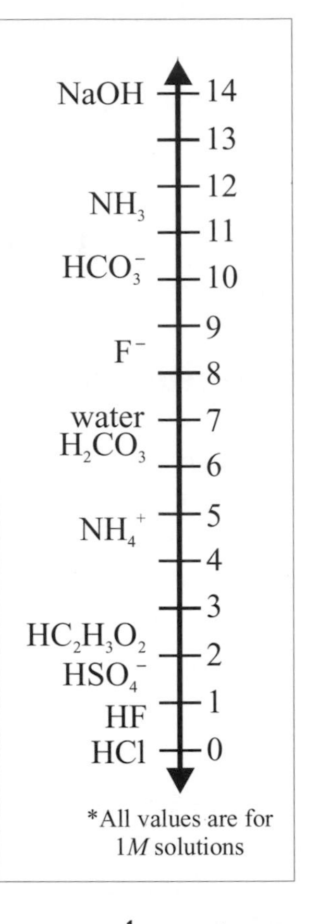

*All values are for 1*M* solutions

The scale for pH generally runs from 0 to 14, but since any H^+ concentration is possible, any pH value is possible. At 25 °C, a pH of 7 is neutral; a lower pH is acidic and a higher pH is basic. Each point on the pH scale corresponds to a tenfold difference in hydrogen ion concentration. An acid with a pH of 2 produces 10 times as many hydrogen ions as an acid with a pH of 3, and 100 times as many as an acid with a pH of 4.

You must understand some very basic ideas about logarithms for the MCAT. pH uses the base 10 logarithm. The base 10 logarithm is used to solve a problem like:

$$10^x = 3.16$$

The answer is: x = log(3.16). Luckily on the MCAT we don't have to do calculations; instead we estimate. Since 10^0 equals 1, and 10^1 equals 10, in the problem above x must be between 0 and 1. The answer is: x = 0.5. Applying this to acids, if we have a hydrogen ion concentration of 10^{-3}, the log of 10^{-3} is -3, and the negative log is positive 3. Thus the pH is 3. If we have a hydrogen ion concentration of a little more than 3, say 4×10^{-3}, then the solution is a little more acidic and the pH is slightly lower: 2.4.

$\log(10^1) = \log(10) = 1$

$\log(3.16) = 0.5$

$\log(10^0) = \log(1) = 0$

Notice that 4×10^{-3} is not as large a number as 10^{-2}, so the pH is lower than 3 but not quite 2. On the MCAT you <u>must</u> be able to estimate pH values as shown in this paragraph.

The second and last thing you should know about logarithms is:

$$\log(AB) = \log(A) + \log(B)$$

This is easily verifiable: $\log(10^2) = 2$; $\log(10^3) = 3$; $\log(10^2\times10^3) = 5$.

One more thing. Many reactions in living cells involve the transfer of a proton. The rate of such reactions depends upon the concentration of H^+ ions or the pH.

From the definitions of an acid, it must be clear that, if there is an acid in a reaction, there must also be a base; you can't have a proton donated without something to accept it. We can write a hypothetical acid-base reaction in aqueous solution as follows:

$$HA + H_2O \rightarrow A^- + H_3O^+$$

Here, HA is the acid, and, since water accepts the proton, water is the base. If we look at the reverse reaction, the hydronium ion donates a proton to A^-, making the hydronium ion the acid and A^- the base. To avoid confusion, we refer to the reactants as the acid and base, and the products as the **conjugate acid** and **conjugate base**. Thus, in every reaction the acid has its conjugate base, and the base has its conjugate acid. Deciding which form is the conjugate simply depends upon in which direction you happen to be viewing the reaction.

In other words, it is correct to say either: "HA is the conjugate acid of base A^-"; or "A^- is the conjugate base of acid HA." You must be able to identify conjugates on the MCAT. You should also know that the stronger the acid, the weaker its conjugate base, and the stronger the base, the weaker its conjugate acid.

Some substances act as either an acid or a base, depending upon their environment. They are called **amphoteric**. Water is a good example. In the reaction above, water acts as a base accepting a proton. Water can also act like an acid by donating a proton.

For the MCAT, you need to recognize the strong acids and bases in Table 6-1.

Strong Acids		Strong Bases	
hydroiodic acid	HI	sodium hydroxide	NaOH
hydrobromic acid	HBr	potassium hydroxide	KOH
hydrochloric acid	HCl	amide ion	NH_2^-
nitric acid	HNO_3	hydride ion	H^-
perchloric acid	$HClO_4$	calcium hydroxide	$Ca(OH)_2$
chloric acid	$HClO_3$	sodium oxide	Na_2O
sulfuric acid	H_2SO_4	calcium oxide	CaO

Table 6-1

By the way, when we say "strong acid" in inorganic chemistry, we mean an acid that is stronger than H_3O^+. A strong base is stronger than OH^-. For MCAT purposes, we assume that a strong acid or base completely dissociates in water.

Some acids can donate more than one proton. These acids are called **polyprotic acids**. An acid that can donate just two protons can also be called a **diprotic acid**. The second proton donated by a polyprotic acid is usually so weak that its effect on the pH is negligible. On the MCAT the second proton can almost always be ignored. (The rule of thumb is that if the K_a values differ by more than 10^3, the second proton can be ignored.) For instance, the second proton from H_2SO_4 is a strong acid; yet, except with dilute concentrations (concentrations less than 1 M), it has a negligible effect on the pH of H_2SO_4 solution. This is because H_2SO_4 is so much stronger then HSO_4^-.

Factors Determining Acid Strength

There are three factors that determine whether or not a molecule containing a hydrogen will release its hydrogen into solution, and thus act as an acid: 1) the strength of the bond holding the hydrogen to the molecule; 2) the polarity of the bond; and 3) the stability of the conjugate base.

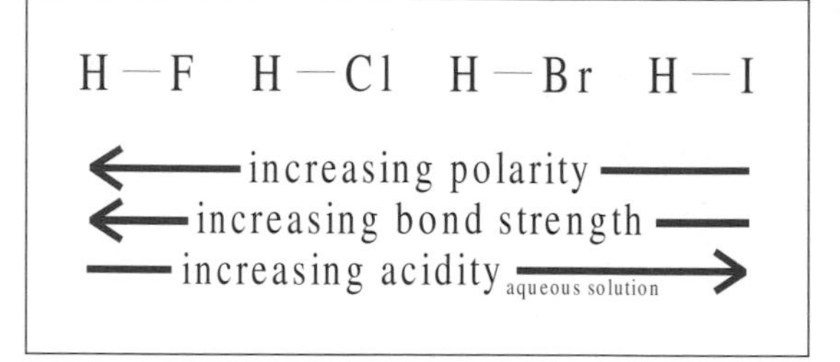

If we examine the C–H bond in methane, which has extremely low acidity, it is slightly weaker than the H–Cl bond in hydrochloric acid. However, the H–Cl bond is much more polar, and therefore is more easily removed in aqueous solution. HCl is more acidic than methane. On the other hand, a comparison of the bond strengths and polarities of the hydrogen halides shows that, although the H–F bond is the most polar, it is also the strongest bond. In addition, the small size of the fluorine ion concentrates the negative charge and adds to its instability. In this case, the bond strength and conjugate instability outweigh the polarity, and HF is the weakest of the hydrogen halide acids.

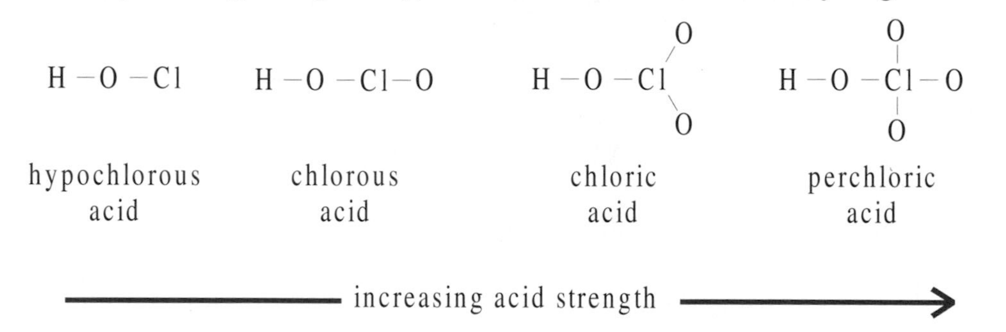

Keeping conjugate stability in mind, if we examine the oxyacids, we see that the electronegative oxygens draw electrons to one side of the bond, increasing polarity. The oxygens in the conjugate of an oxyacid can share the negative charge spreading it over a larger area and stabilizing the conjugate base. In similar oxyacids, the molecule with the most oxygens makes the strongest acid. Another way to look at this phenomenon is that the acidity increases with the oxidation number of the central atom.

By the way, the percent dissociation of an acid decreases with acidity. This means that strong acids dissociate less in very concentrated solutions.

Hydrides

Binary compounds (compounds with only two elements) containing hydrogen are called hydrides. Hydrides can be basic, acidic, or neutral. On the periodic table, the basic hydrides are to the left, and the acidic hydrides are to the right. For instance, NaH is basic; H_2S is acidic. Following this trend, metal hydrides are either basic or neutral, while nonmetal hydrides are acidic or neutral. (Ammonia, NH_3, is an exception to this rule.) The acidity of nonmetal hydrides tends to increase going down the periodic table. $H_2O < H_2S < H_2Se < H_2Te$

Questions 76 through 80 are **NOT** based on a descriptive passage.

76. Ammonia reacts with water to form the ammonium ion and hydroxide ion.

$$NH_3 + H_2O \rightarrow NH_4^+ + OH^-$$

According to the Bronsted-Lowry definition of acids and bases, what is the conjugate acid of ammonia?

A. NH_3
B. NH_4^+
C. OH^-
D. H^+

77. By definition, a Lewis base:

A. donates a proton.
B. accepts a proton.
C. donates a pair of electrons.
D. accepts a pair of electrons.

78. Which of the following is the strongest base in aqueous solution?

A. Cl^-
B. NH_4^+
C. F^-
D. Br^-

79. Which of the following is amphoteric?

A. an amino acid
B. H_2SO_4
C. NaOH
D. HF

80. The addition of an electron withdrawing group to the alpha carbon of a carboxylic acid will:

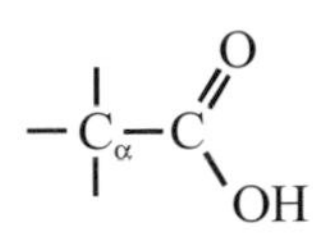

A. increase the acidity of the proton by making the O-H bond more polar.
B. increase the acidity of the proton by making the O-H bond stronger.
C. decrease the acidity of the proton by making the O-H bond more polar.
D. decrease the acidity of the proton by stabilizing the conjugate base.

Answers to Questions 76-80

76. B is correct. The conjugate acid is the molecule after it accepts a proton.

77. C is correct. By definition, a Lewis base donates a pair of electrons.

78. C is correct. NH_4^+ is an acid. The strongest base is the conjugate of the weakest acid.

79. A is correct. An amino acid can act as an acid or a base depending upon the pH. Although the conjugate base of sulfuric acid is amphoteric, sulfuric acid cannot accept a proton and is not amphoteric.

80. A is correct. The electron withdrawing group will further polarize the O-H bond, and polarization increases acidity in aqueous solution.

Equilibrium Constants for Acid-Base Reactions

Pure water reacts with itself to form hydronium and hydroxide ions as follows:

$$H_2O + H_2O \rightarrow H_3O^+ + OH^-$$

This is called the **autoionization of water**. K_w is the equilibrium constant for this reaction.

$$K_w = [H^+][OH^-]$$

(For convenience, we have substituted H^+ for H_3O^+.) At 25 °C the equilibrium of this reaction lies far to the left:

$$K_w = 10^{-14}$$

In a neutral aqueous solutions at 25 °C, the H^+ concentration and the OH^- concentration are equal at 10^{-7} mol L^{-1}. The pH of the solution is found by taking the negative log of the hydrogen ion concentration, which is: $-\log[10^{-7}] = 7$. An acid or base will change the concentrations of both H^+ and OH^-, but K_w will remain 10^{-14} at 25 °C. For example, in a solution with a pH of 2, the ion concentrations will be: $[H^+] = 10^{-2}$ mol L^{-1} and $[OH^-] = 10^{-12}$ mol L^{-1}. Using the p(x) function and the rule: log(AB) = log(A) + log(B), we can put this relationship into a simple equation:

$$\mathrm{pH} + \mathrm{pOH} = \mathrm{p}K_w$$

For an aqueous solution at 25 °C:

$$\mathrm{pH} + \mathrm{pOH} = 14$$

An acid will have its own equilibrium constant in water, called the **acid dissociation constant K_a**. If we use our hypothetical acid-base reaction: $HA + H_2O \rightarrow H_3O^+ + A^-$, then the acid dissociation constant for the acid HA is:

$$K_a = \frac{[H^+][A^-]}{[HA]}$$

Corresponding to every K_a, there is a $\underline{\boldsymbol{K_b}}$. The K_b is the equilibrium constant for the reaction of the conjugate base with water. For the conjugate base A^- in our hypothetical reaction, the reaction is:

$$A^- + H_2O \rightarrow OH^- + HA$$

and the K_b is:

$$\boldsymbol{K_b = \frac{[OH^-][HA]}{[A^-]}}$$

Notice that the reaction for K_b is the reaction of the conjugate base and water, and it is <u>not</u> the reverse of the reaction for K_a. Notice also that the product of the two constants is K_w.

$$K_aK_b = \frac{[H^+][\cancel{A^-}]}{[\cancel{HA}]} \times \frac{[OH^-][\cancel{HA}]}{[\cancel{A^-}]} = [H^+][OH^-] = K_w$$

$$\boldsymbol{K_aK_b = K_w}$$

Using the p(x) function and the rule: log(AB) = log(A) + log(B), this formula can also be written as:

$$\mathbf{p}\boldsymbol{K_a} + \mathbf{p}\boldsymbol{K_b} = \mathbf{p}\boldsymbol{K_w}$$

At 25 °C:

$$\mathbf{p}\boldsymbol{K_a} + \mathbf{p}\boldsymbol{K_b} = \mathbf{14}$$

It may seem like there are a lot of equations to memorize here, but it is really very simple. First, all equilibrium constants are derived from the law of mass action. They are all products over reactants, where only products and reactants that are in the same phase are used. Once you know one K, you should know all of them. The subscript on the constant is supposed to make things less complicated not more complicated.

Second, memorize that $K_w = 10^{-14}$ at 25 °C.

Third, remember the log rule, log(AB) = log(A) + log(B), and you can derive any of the equations above.

Notice that the larger the K_a, the stronger the acid, but the smaller the pK_a, the stronger the acid.

Finding the pH

Very strong acids and bases will dissociate almost completely. This means that the HA or BOH concentration (for the acid and base respectively) will be nearly zero. Since division by zero is impossible, for such acids and bases, there is no K_a or K_b. Surprisingly, this fact makes it easier to find the pH of strong acid and strong base solutions. Since the entire concentration of acid or base is assumed to dissociate, the concentration of H^+ or OH^- is the same as the original concentration of acid or base. For instance, a 0.01 molar solution of HCl will have 0.01 mol L^{-1} of H^+ ions. Since $0.01 = 10^{-2}$, and $-\log(10^{-2}) = 2$, the pH of the solution will be 2. Likewise, in a 0.01 molar solution of NaOH, we will have 0.01 mol L^{-1} of OH^- ions. (Be careful here!) The pOH will equal 2 so the pH will equal 12. You can avoid a mistake here by remembering that an acid has a pH below 7 and a base has a pH above 7.

Weak acids and bases can be a little trickier. Doing a sample problem is the best way to learn. For example, in order to find the pH of a 0.01 molar solution of HCN, we do the following:

1) Set up the equilibrium equation:

$$K_a = \frac{[H^+][CN^-]}{[HCN]} = 6.2\text{x}10^{-10}$$

2) If we add 0.01 moles of HCN to one liter of pure water, then 'x' amount of that HCN will dissociate. Thus, we will have 'x' mol L^{-1} of H^+ ions and 'x' mol L^{-1} of CN^- ions. The concentration of undissociated HCN will be whatever is left, or '$0.01 - x$'. Plugging in these numbers to the equation above, we have:

$$\frac{[x][x]}{[0.01 - x]} = 6.2\text{x}10^{-10}$$

3) If we solve for x, we have a quadratic equation. Forget it! You don't need this for the MCAT. We make an assumption that x is less than 5% of 0.01, and we will check it when we are done. Throwing out the x in the denominator, we have:

$$\frac{[x][x]}{[0.01]} \approx 6.2\text{x}10^{-10}$$

Thus, x is approximately $2.5\text{x}10^{-6}$. This is much smaller than 0.01, so our assumption was valid. 'x' is the concentration of H^+ ions. The pH of the solution is between 5 and 6. This is close enough for the MCAT. [$-\log(2.5\text{x}10^{-6}) = 5.6$] Just to make sure, we ask ourselves, "Is 5.6 a reasonable pH for a dilute weak acid?" The answer is yes.

For a weak base, the process is the same, except that we use K_b, and we arrive at the pOH. Subtract the pOH from 14 to find the pH. This step is often forgotten. If we ask ourselves, "Is this pH reasonable for a weak base?" we won't forget this step.

Salts

Salts are ionic compounds that dissociate in water. Often, when salts dissociate, they create acidic or basic conditions. The pH of a salt solution can be predicted qualitatively by comparing the conjugates of the respective ions. Just keep in mind that strong acids have weak conjugate bases and strong bases have weak conjugate acids.

Na^+ and Cl^- are the conjugates of NaOH and HCl respectively, so, as a salt, NaCl produces a neutral solution. NH_4NO_3 is composed of the conjugates of the base NH_3 and the strong acid HNO_3 respectively. Thus, NH_4^+ is acidic and NO_3^- is neutral. As a salt, NH_4NO_3 is weakly acidic.

When considering salts, remember, all cations, except those of the alkali metals and the heavier alkaline earth metals (Ca^{2+}, Sr^{2+}, and Ba^{2+}), act as weak Lewis acids in aqueous solutions.

Questions 81 through 85 are **NOT** based on a descriptive passage.

81. Which of the following is the K_b for the conjugate base of carbonic acid?

A. $\frac{[H_2CO_3]}{[H^+][HCO_3^-]}$

B. $\frac{[OH^-][HCO_3^-]}{[H_2CO_3]}$

C. $\frac{[H^+][H_2CO_3]}{[HCO_3^-]}$

D. $\frac{[OH^-][H_2CO_3]}{[HCO_3^-]}$

82. An aqueous solution of 0.1 *M* HBr has a pH of:

A. 0
B. 1
C. 2
D. 14

83. Carbonic acid has a K_a of 4.3×10^{-7}. What is the pH when 1 mole of $NaHCO_3$ is dissolved in 1 liter of water?

A. 3.2
B. 3.8
C. 10.2
D. 12.5

84. Stomach acid has a pH of approximately 2. Sour milk has a pH of 6. Stomach acid is:

A. 3 times as acidic as sour milk.
B. 4 times as acidic as sour milk.
C. 100 times as acidic as sour milk.
D. 10,000 times as acidic as sour milk.

85. Which of the following salts is the most basic?

A. NaI
B. $NaNO_3$
C. NH_4Cl
D. KOH

Answers to Questions 81-85

81. D is correct. K_b is the reaction of the conjugate base with water.

82. B is correct. HBr dissociates completely, so the concentration of H^+ ions will be equal to the concentration of solution. The $-\log(0.1) = 1$.

83. C is correct. The K_b for $NaHCO_3$ is $K_w/K_a \approx \frac{1}{4} \times 10^{-7}$. We can set up the equilibrium expression:

$$K_b = \frac{[OH^-][H_2CO_3]}{[HCO_3^-]}$$

$$0.25 \times 10^{-7} = \frac{[x][x]}{[1 - x]}$$

This x is insignificant.

$$2.5 \times 10^{-8} = x^2$$

$$1.5 \times 10^{-4} = x$$

Thus, the pOH = between 3 and 4. Subtracting from 14, the pH = between 10 and 11.

84. D is correct. Each unit of pH is a tenfold increase of acidity.

85. D is correct. You should recognize potassium hydroxide as a strong base.

Titrations

A **titration** is the drop-by-drop mixing of an acid and a base. The changing pH of the mixture is represented graphically as a sigmoidal curve. Below is the **titration curve** of a strong acid titrated by a strong base.

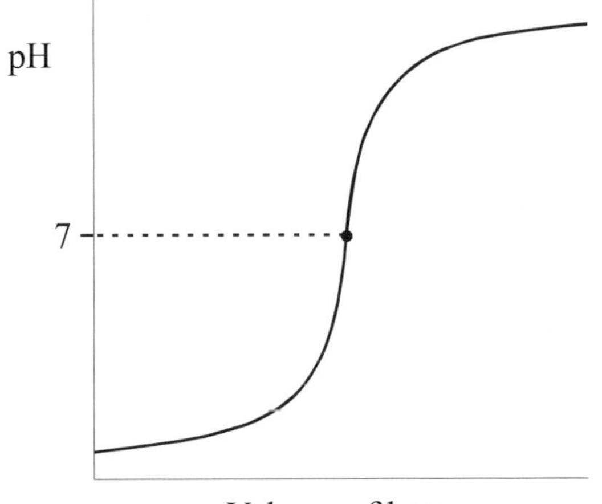

Notice the portion of the graph that most nearly approximates a vertical line. The midpoint of this line is called the **equivalence point** or the **stoichiometric point**. The equivalence point for a monoprotic acid is the point in the titration when there are equal equivalents of acid and base in solution. (An *equivalent* is the mass of acid or base necessary to produce or consume one mole of protons.) For instance, since there is a one to one correspondence between HCl with NaOH, the equivalence point for a titration of HCl with NaOH will be reached when the same number of moles of HCl and NaOH exist in solution. This is not necessarily when they are at equal volumes. If the concentrations differ (and they probably will) the equivalence point will not be where the volumes are equal.

For equally strong acid-base titrations, the equivalence point will usually be at pH 7. (**Warning!** for a diprotic acid whose conjugate base is a strong acid, like H_2SO_4, this is not the case.)

The above graph is for the titration of a strong acid with a strong base. In other words, we are slowly adding base to an acid. This is clear because we start with a very low pH and finish with a very high pH. For a titration of a strong base with a strong acid, we would simply invert the graph.

More Titrations and Buffered Solutions

The titration of a weak acid with a strong base (or a weak base with a strong acid) looks slightly different than the curve above, and is shown below. The equivalence point is also not as predictable. Of course, if the base is stronger than the acid, the equivalence point will be above 7, and if the acid is stronger than the base, the equivalence point will be below 7.

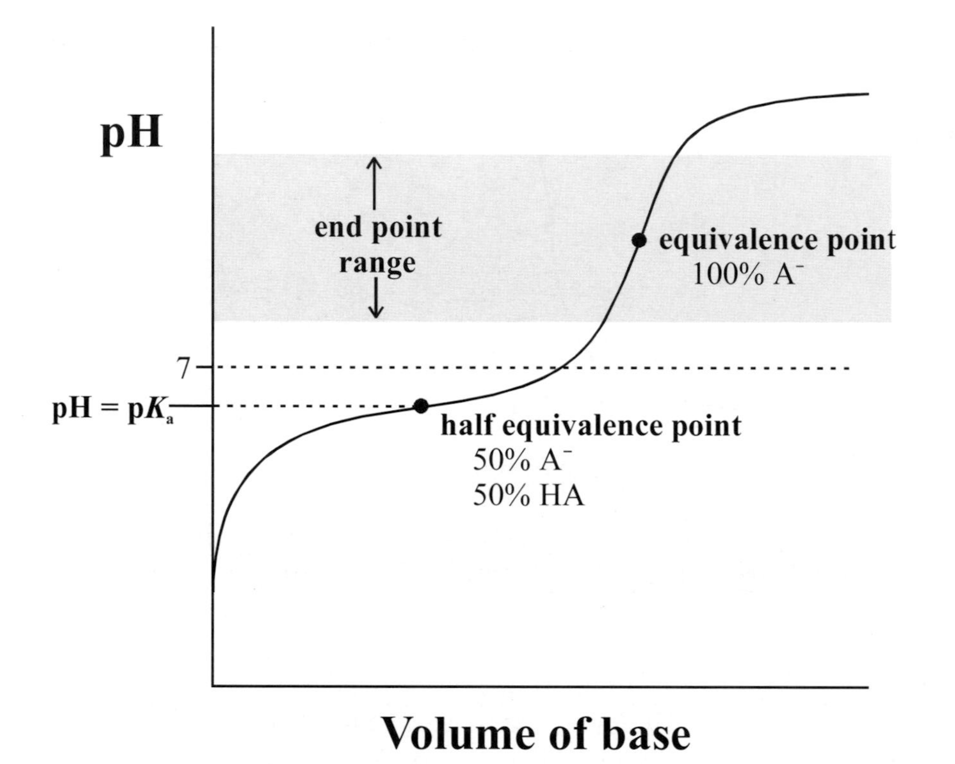

Notice the **half equivalence point**. This is probably more germane to the MCAT than the equivalence point. The half equivalence point is the point where exactly one half of the acid has been neutralized by the base. In other words, the concentration of the acid is equal to the concentration of its conjugate base. Notice that the half equivalence point occurs at the midpoint of the section of the graph that most represents a horizontal line. This is the spot where we could add the largest amount of base or acid with the least amount of change in pH. Such a solution is considered to be **buffered**. The half equivalence point shows the point in the titration where the solution is the most well buffered.

Notice also that, at the half equivalence point, the pH of the solution is equal to the pK_a of the acid. This is predicted by the **Henderson-Hasselbalch equation**:

$$\mathbf{pH = p}K_a + \mathbf{log}\,\frac{[\mathrm{A}^-]}{[\mathrm{HA}]}$$

Recall that log(1) = 0; thus when $[\mathrm{A}^-] = [\mathrm{HA}]$, pH = p$K_a$.

The Henderson-Hasselbalch equation is simply a form of the equilibrium expression for K_a:

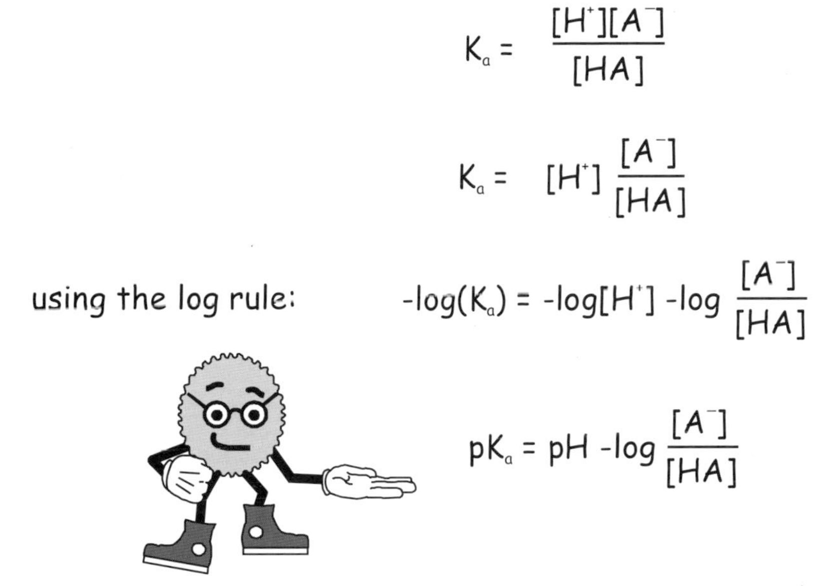

$$K_a = \frac{[\mathrm{H}^+][\mathrm{A}^-]}{[\mathrm{HA}]}$$

$$K_a = [\mathrm{H}^+]\,\frac{[\mathrm{A}^-]}{[\mathrm{HA}]}$$

using the log rule: $$-\log(K_a) = -\log[\mathrm{H}^+] - \log\frac{[\mathrm{A}^-]}{[\mathrm{HA}]}$$

$$\mathrm{p}K_a = \mathrm{pH} - \log\frac{[\mathrm{A}^-]}{[\mathrm{HA}]}$$

There is no need to memorize it, since it is so easy and so quick to derive it.

If we were to make a buffer solution, we would start with an acid whose pK_a is closest to the pH at which we want to buffer our solution. Next we would mix equal amounts of that acid with its conjugate base. We would want the concentration of our buffer solution to greatly exceed the concentration of outside acid or base affecting our solution. So, a buffer solution is made from equal and copious amounts of a weak acid and its conjugate base.

It appears from the Henderson-Hasselbalch equation that we could add an infinite amount of water to a buffered solution with no change in pH. Of course, this is ridiculous. Will adding Lake Tahoe to a beaker of buffered solution change the pH of that solution? The Henderson-Hasselbalch equation does not allow for *ion pairing*. (Ion pairing is when oppositely charged ions in solution bond momentarily to form a single particle.) Water will generally act like a base in acidic solution and an acid in basic solution. If you add a base or water to an acidic, buffered solution, it is clear from the titration curve that the pH will increase. It just won't increase as rapidly as other solutions less well buffered. However, a question on the MCAT is more likely to consider the ideal circumstance where adding a small amount of water to an ideally dilute, buffered solution will have no effect on the pH.

Warning! You cannot typically use the Henderson-Hasselbalch equation to find the pH at the equivalence point. Instead, you must use the K_b of the conjugate base. You can find the K_b from the K_a and the K_w. The concentration of the conjugate base at the equivalence point is equal to the number of moles of acid divided by the volume of acid plus the volume of base used to titrate. Don't forget to consider the volume of base used to titrate. Unless the base has no volume, the concentration of the conjugate at the equivalence point will not be equal to the original concentration of the acid. The pH at the equivalence point involves much more calculation than the pH at the half equivalence point. For this reason, it is more likely that the MCAT will ask about the pH at the half equivalence point.

Finding the pH at the equivalence point is a good exercise, but you won't have to do it on the MCAT. Here are the steps:

Use K_a and K_w to find the K_b

$$K_b = \frac{K_w}{K_a}$$

Set up the K_b equilibrium expression.

$$K_b = \frac{[OH^-][HA]}{[A^-]}$$

Solve for the OH^- concentration, and find the pOH.

Subtract the pOH from 14 to find the pH.

$$14 - pOH = pH$$

Indicators and the End Point

To find the equivalence point, a chemical called an **indicator** is used. (A pH meter can also be used.) The indicator is usually a weak acid whose conjugate base is a different color. We can designate an indicator as HIn, where In^- represents the conjugate base. In order for the human eye to detect a color change, the new form of the indicator must reach 1/10 the concentration of the original form. For example, if we titrate an acid with a base, we add a small amount of indicator to our acid. (We add only a small amount because we don't want the indicator to affect the pH.) At the initial low pH, the HIn form of the indicator predominates. As we titrate, and the pH increases, the In^- form of the indicator also increases. When the In^- concentration reaches 1/10 of the HIn concentration, a color change can be detected by the human eye. If we titrate a base with an acid, the process works in reverse. Thus, the pH of the color change depends upon the direction of the titration. The pH values of the two points of color change give the **range** of an indicator. An indicator's range can be predicted by using the Henderson-Hasselbalch equation as follows:

$$pH = pK_a + \log \frac{[In^-]}{[HIn]}$$

lower range of color change ==> $pH = pK_a + \log \frac{1}{10}$ ==> $pH = pK_a - 1$

upper range of color change ==> $pH = pK_a + \log \frac{10}{1}$ ==> $pH = pK_a + 1$

The point where the indicator changes color is called the **endpoint**. Do not confuse the equivalence point with the end point. We usually choose an indicator whose range will cover the equivalence point.

You can also monitor the pH with a pH meter. A pH meter is a concentration cell comparing the voltage difference between different concentrations of H^+. (See Chemistry Lecture 7 for concentration cells.)

You don't need to memorize this stuff about indicators, but its useful to understand.

By the way, you can remember that the end point is where the indicator changes color by spelling indicator as

Endicator.

Since we established that the Henderson-Hasselbalch equation is not useful to find the pH of the equivalence point, how can it be useful to find an indicator range that will include the equivalence point?

The answer is that we are using the indicator concentrations in the Henderson-Hasselbalch equation, and the indicator never reaches its equivalence point in the titration. The indicator ions do not approach zero concentration near the color change range.

Polyprotic Titrations

Titrations of polyprotic acids will have more than one equivalence point and more than one half equivalence point. For the MCAT, assume that the first proton completely dissociates before the second proton begins to dissociate. (This assumption is only acceptable if the second proton is a much weaker acid than the first.) Thus we have a titration curve like the one shown below.

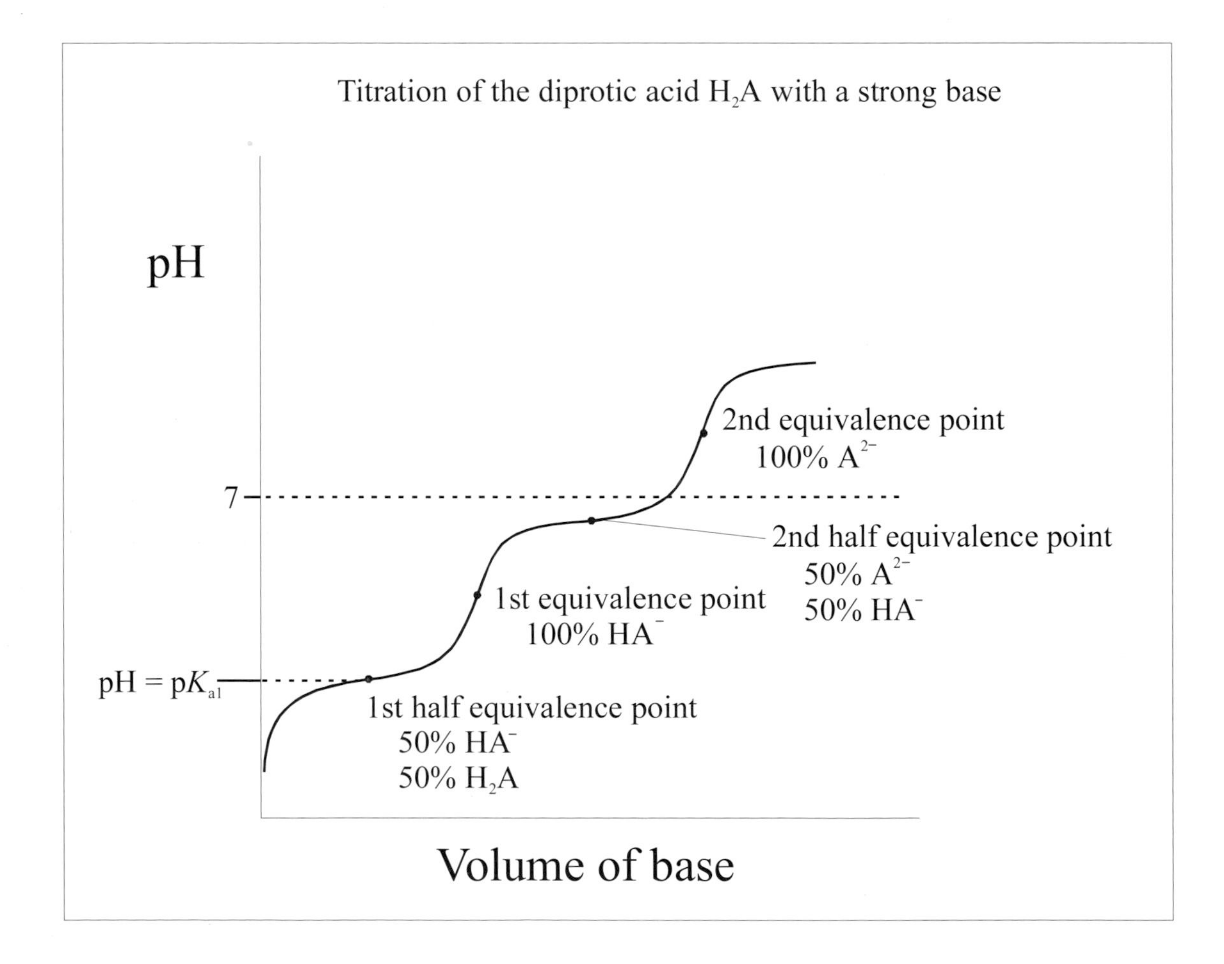

Questions 86 through 90 are **NOT** based on a descriptive passage.

86. The titration curve below represents the titration of:

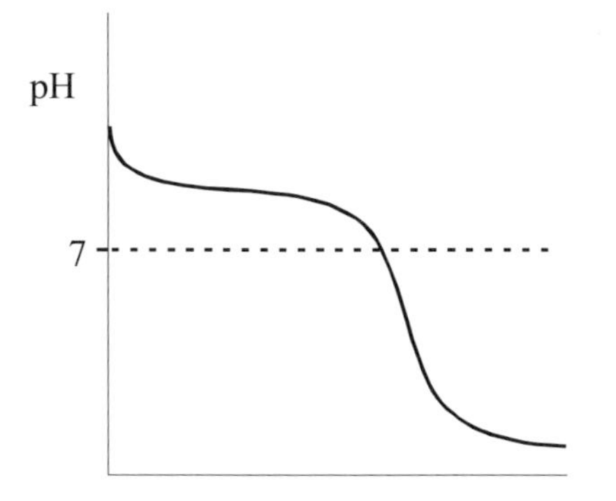

A. a strong acid with a weak base.
B. a strong base with a weak acid.
C. a weak acid with a strong base.
D. a weak base with a strong acid.

87. The following is a list of acid dissociation constants for 4 acids.

	K_a
Acid 1	1.2×10^{-7}
Acid 2	8.3×10^{-7}
Acid 3	3.3×10^{-6}
Acid 4	6.1×10^{-5}

Which acid should be used to manufacture a buffer at a pH of 6.1?

A. Acid 1
B. Acid 2
C. Acid 3
D. Acid 4

88. If the expected equivalence point for a titration is at a pH of 8.2, which of the following would be the best indicator for the titration?

Indicator	K_a
phenolphthalein	1.0×10^{-8}
bromthymol blue	7.9×10^{-8}
methyl orange	3.2×10^{-4}
methyl violet	1.4×10^{-3}

A. phenolphthalein
B. bromthymol blue
C. methyl orange
D. methyl violet

89. On the titration curve of the H_2CO_3 pictured below, at which of the following points is the concentration of HCO_3^- the greatest?

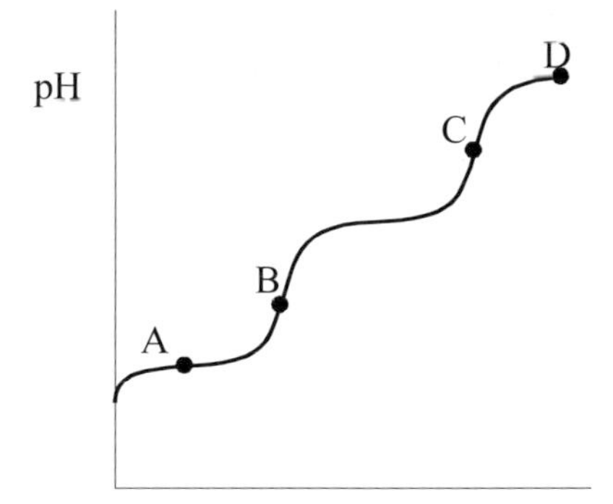

A. point A
B. point B
C. point C
D. point D

90. Which of the following is the equivalence point when the weak acid, acetic acid, is titrated with NaOH?

A. 4.3
B. 7
C. 8.7
D. 14

Answers to Questions 86-90

86. D is correct. The pH starts basic so a base is being titrated. It ends very acidic so a strong acid is titrating.

87. B is correct. A buffer is made from equal amounts of an acid and its conjugate. The buffer works best when the pH = pK_a. $-\log(8.3\text{x}10^{-7})$ = between 6 and 7. 8.3 is close to ten, making the pK_a closer to 6.

88. A is correct. An indicator generally changes color within plus or minus one pH point of its pK_a.

89. B is correct. The concentration of the conjugate base of the first acid is the greatest at the first equivalence point.

90. C is correct. The equivalence point of a titration of a weak acid with a strong base will always be greater than 7. It is the same as adding the conjugate base of the acid to pure water. 14 is way too basic. Pure 1 *M* NaOH has a pH of 14.

Lecture 7: Electrochemistry

Oxidation-Reduction

In an oxidation-reduction reaction (called a **redox reaction** for short), electrons are transferred from one atom to another. The atom that loses electrons is **oxidized**; the atom that gains electrons is **reduced**. In order to keep track of the electrons in a redox reaction, you must memorize the **oxidation states** of certain atoms. Oxidation states are the possible charge values that an atom may hold within a molecule. These charges don't truly exist in many cases; it is simply a system to follow the electrons of a redox reaction. Even though they do not represent actual charges, the oxidation states must add up to the charge on the molecule. For instance, the sum of the oxidation states of the atoms in a neutral molecule must equal zero. The oxidation potentials that you must memorize for the MCAT are as follows:

Oxidation State	Atom
0	Atoms in their elemental form
+1	Hydrogen (except when bonded to a metal: then –1.)
–2	Oxygen (except when it is in a peroxide like H_2O_2)

Table 7-1

In general, when in a compound, elements in the following groups have the oxidation states listed in the table below. It is helpful to know this second table but not crucial to the MCAT.

Oxidation State	Group on Periodic Table
+1	Group 1 elements (alkali metals)
+2	Group 2 elements (alkaline earth metals)
–3	Group 15 elements (nitrogen family)
–2	Group 16 elements (oxygen family)
–1	Group 17 elements (halogens)

Table 7-2

The idea is simple; a general guideline for oxidation states is the atom's variance from a noble gas configuration. However, if all atoms had permanent oxidation states, no redox reactions could take place. The oxidation states in the second table are to be used only as a general guideline. When the two tables conflict, the first table is given priority. For example, the oxidation state of nitrogen in NO_3^- is +5 because the –2 on the oxygens have priority and dictate the oxidation state on nitrogen. (Don't forget that the oxidation states for NO_3^- must add up to the 1– charge on the molecule.) The transition metals change oxidation states according to the atoms with which they are bonded. Although each transition metal has only certain oxidation states that it can attain, the MCAT will not require that you memorize these.

It is very unlikely that you will be required by the MCAT to know more oxidation states than the ones given for hydrogen, oxygen, and atoms in their elemental form. To help keep oxidation and reduction straight, just remember:

A typical redox reaction is as follows:

$$2H_2 + O_2 \rightarrow 2H_2O$$

Here oxygen and hydrogen begin in their elemental form, and thus have an oxidation state of zero. Once the water molecule is formed, hydrogen's oxidation state is +1, and oxygen's is –2. In this case, we say that hydrogen has been **oxidized**; hydrogen has lost electrons; its oxidation state has increased from 0 to +1. Oxygen, on the other hand, has been **reduced**; it has gained electrons; its oxidation state has been reduced from 0 to –2. Whenever there is oxidation, there must also be reduction.

Since in any redox reaction an atom is oxidized and an atom is reduced, there is a **reducing agent** (also called the **reductant**) and an **oxidizing agent** (also called the **oxidant**). Because the reducing agent is giving an atom electrons, an atom in the reducing agent must be giving up some of its own electrons. Since an atom in the reducing agent gives up electrons, an atom in the reducing agent is oxidized. The reverse is true for the oxidizing agent. Thus, the reducing agent is the compound containing the atom being oxidized, and the oxidizing agent is the compound containing the atom being reduced. For example, in the following reaction, methane is the reducing agent and oxygen is the oxidizing agent.

Reducing Agent Salty

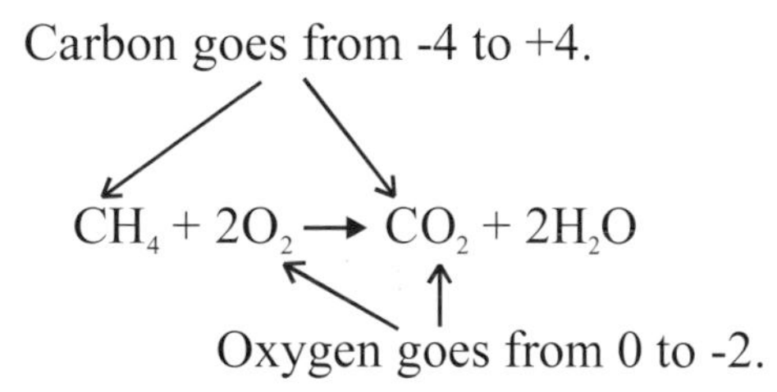

Notice that the reducing agents and oxidizing agents are compounds, not atoms. In a redox reaction, the atom is oxidized or reduced; the compound is the oxidant or reductant. In the reaction:

$$Cd(s) + NiO_2(s) + 2H_2O(l) \rightarrow Cd(OH)_2(s) + Ni(OH)_2(s)$$

Ni is reduced. NiO_2 is the oxidizing agent.

Oxidation-Reduction Titrations

In order to find the molarity of a reducing agent, a sample can be titrated with a strong oxidizing agent. For instance, if we want to know the molarity of Sn^{2+} ions in a solution, we can titrate it with a known concentration of the strong oxidizing agent Ce^{4+}. Sn^{2+} ions oxidize to Sn^{4+}, while Ce^{4+} reduces to Ce^{3+}. Since only one electron is required to reduce Ce^{4+}, and two electrons are given up to oxidize Sn^{2+}, two Ce^{4+} ions are reduced for every Sn^{2+} ion oxidized. Thus, we know that the number of moles of Ce^{4+} required to reach the equivalence point is twice the number of moles of Sn^{2+} in solution. Instead of measuring the pH, we measure the voltage compared to a standard solution.

Knowledge of oxidation-reduction titrations is not required for the MCAT. However, it is possible that there will be a passage which explains them. They are included here just so you won't be shocked if you see one in a passage.

Questions 91 through 95 are **NOT** based on a descriptive passage.

91. What is the oxidation state of sulfur in HSO_4^-?

A. –2
B. +3
C. +6
D. +7

92. Which of the following statements is true concerning the reaction:

$$2Al_2O_3 + 3C \rightarrow 4Al + 3CO_2$$

A. Both aluminum and carbon are reduced.
B. Both aluminum and carbon are oxidized.
C. Aluminum is reduced and carbon is oxidized.
D. Carbon is reduced and aluminum is oxidized.

93. What is the reducing agent in the following reaction:

$$HCl + Zn \rightarrow ZnCl_2 + H_2$$

A. Zn
B. Zn^{2+}
C. H^+
D. Cl^-

94. The first step in producing pure lead from galena (PbS) is as follows:

$$2PbS(s) + 3O_2(g) \rightarrow 2PbO(s) + 2SO_2(g)$$

All of the following are true concerning this reaction EXCEPT:

A. Both lead and sulfur are oxidized.
B. Oxygen is the oxidizing agent.
C. Lead sulfide is the reducing agent.
D. Lead is neither oxidized nor reduced.

95. All of the following are true concerning oxidation-reduction reactions EXCEPT:

A. An atom in the reducing agent is always oxidized.
B. If reduction takes place, so must oxidation.
C. An atom in the oxidizing agent gains electrons.
D. If an atom of the reductant loses two electrons, an atom of the oxidant gains two electrons.

Answers to Questions 91-95

91. C is correct. Each oxygen has an oxidation state of –2, and hydrogen has an oxidation state of +1. In order for the ion to have a 1– charge, the sulfur must have a +6 oxidation state. (Notice that oxidation states are given as +*n*, and actual charges are given as *n*+.)

92. C is correct. Aluminum begins as +3 and ends as 0, while carbon begins as 0 and ends as +4.

93. A is correct. The Zn is oxidized from an oxidation state of 0 to +2. Thus, it is the reducing agent.

94. A is correct. Both A and D cannot be true, so the answer must be A or D. The trickiest part of this problem is to know that lead is comfortable at 2+ and sulfur, being in the oxygen family, is comfortable at –2; thus these are their oxidation states when they are together. But when they are with oxygen, the –2 of the oxygen rules.

95. D is correct. An example of where this is false is:

$$2HCl + Zn \rightarrow ZnCl_2 + H_2$$

Here each atom of the reducing agent, zinc, loses two electrons, and the hydrogen atom of the oxidizing agent, HCl, gains one electron. Of course, there must be two hydrogens for each zinc.

Potentials

Since in a redox reaction electrons are transferred, and since electrons have charge, there is an electric potential $\mathscr{E}$ associated with any redox reaction. The potentials for the oxidation component and reduction component of a reaction can be measured separately. Each component is called a **half reaction**. Of course, no half reaction will occur by itself; any reduction half reaction must be accompanied by an oxidation half reaction. The potential for any given atom to be reduced to a given state, is always the same, regardless of the reaction. Thus there is only one potential for any given half reaction. Since the reverse of a reduction half reaction is an oxidation half reaction, it would be redundant to list potentials for both the oxidation and reduction half reactions. Therefore, half reaction potentials are usually listed as **reduction potentials**. To find the oxidation potential for the reverse half reaction, the sign of the reduction potential is reversed. Below is a list of some common reduction potentials.

Standard reduction potentials at 25 °C	
Half reaction	**Potential $\mathscr{E}^o$**
$Au^{3+}(aq) + 3e^- \rightarrow Au(s)$	1.50
$O_2(g) + 4H^+(aq) + 4e^- \rightarrow H_2O(l)$	1.23
$Pt^{2+}(aq) + 2e^- \rightarrow Pt(s)$	1.2
$Ag^{2+}(aq) + 2e^- \rightarrow Ag(s)$	0.80
$Hg^{2+}(aq) + 2e^- \rightarrow Hg(l)$	0.80
$Cu^{+}(aq) + e^- \rightarrow Cu(s)$	0.52
$Cu^{2+}(aq) + 2e^- \rightarrow Cu(s)$	0.34
$\mathbf{2H^{+}(aq) + 2e^- \rightarrow H_2(g)}$	**0.00**
$Ni^{2+}(aq) + 2e^- \rightarrow Ni(s)$	–0.23
$Fe^{3+}(aq) + 3e^- \rightarrow Fe(s)$	–0.036
$Fe^{2+}(aq) + 2e^- \rightarrow Fe(s)$	–0.44
$Zn^{2+}(aq) + 2e^- \rightarrow Zn(s)$	–0.76
$H_2O(l) + 2e^- \rightarrow H_2(g) + 2OH^-(aq)$	–0.83

Table 7-3

Recall from physics that electric potential has no absolute value. Thus the values in the table above are assigned based upon the arbitrary assignment of a zero value to the half reaction:

$$\mathbf{2H^+ + 2e^- \rightarrow H_2 \quad \mathscr{E}^o = 0.00\ V}$$

This is the only reduction potential that you need to memorize.

An example of an oxidation potential taken from the table above would be:

$$Ag(s) \rightarrow Ag^{2+}(aq) + e^- \quad \mathscr{E}^o = -0.80\ V$$

Notice that, except for nickel, the metals used to make coins have negative oxidation potentials. In other words, unlike most metals, platinum, gold, silver, mercury, and copper do not oxidize (dissolve) spontaneously in the presence of aqueous H^+.

Also notice that Table 7-3 gives us the reduction potential for Ag^{2+}(*aq*) and the oxidation potential for Ag(*s*). (**Warning:** The table does not give us the oxidation potential for Ag^{2+}.) The strongest oxidizing agent is shown on the upper left hand side of a reduction table. The strongest reducing agent is shown on the lower right hand side of a reduction table. Notice that water is both a poor oxidizing agent and a poor reducing agent.

Finally, notice that the second half reaction in Table 7-3 is part of the final reaction in aerobic respiration where oxygen accepts electrons to form water.

Reduction potential is an *intensive* property (see Physics Lecture 5). Thus if we wish to find the potential of the following ionic reaction:

$$2Au^{3+} + 3Cu \rightarrow 3Cu^{2+} + 2Au$$

we can separate the reaction into its two half reactions and add the half reaction potentials:

$$\begin{aligned} 2(Au^{3+} + 3e^- \rightarrow Au) \quad \mathscr{E}^o &= 1.50\text{ V} \\ 3(Cu \rightarrow Cu^{2+} + 2e^-) \quad \mathscr{E}^o &= \underline{-0.34\text{ V}} \\ &= 1.16\text{ V} \end{aligned}$$

Warning: Do not multiply the half reaction potential by the number of times it occurs.

Balancing Redox Reactions

Balancing redox reactions can be tricky. When you have trouble, follow the steps below to balance a redox reaction that occurs in acidic solution.

1. Divide the reaction into its corresponding half reactions.
2. Balance the elements other than H and O.
3. Add H_2O to one side until the O atoms are balanced.
4. Add H^+ to one side until the H atoms are balanced.
5. Add e^- to one side until the charge is balanced.
6. Multiply each half reaction by an integer so that the electrons balance.
7. Add the two half reactions and simplify.

For redox reactions occurring in basic solution, follow the same steps, then neutralize the H^+ ions by adding the same number of OH^- ions to both sides of the reaction.

Galvanic Cell

Although each redox reaction has a potential associated with it, if we simply mix the chemicals and allow the reaction to proceed, we cannot take advantage of this potential. In order to take advantage of this potential, we separate the two half reactions, and allow the electron flow between them to take place through a conductor. (**Warning:** As stated earlier, a half reaction will not proceed by itself.) Since the flowing electrons would create a build up of negative charge in one of the solutions, we add a **salt bridge** to allow the charge to remain neutral. The salt bridge allows positive ions to flow toward the cathode and negative ions, toward the anode. The ions in the salt bridge should not react with other ions or the electrodes. The entire apparatus just described is a **galvanic cell** (also called a **voltaic cell**).

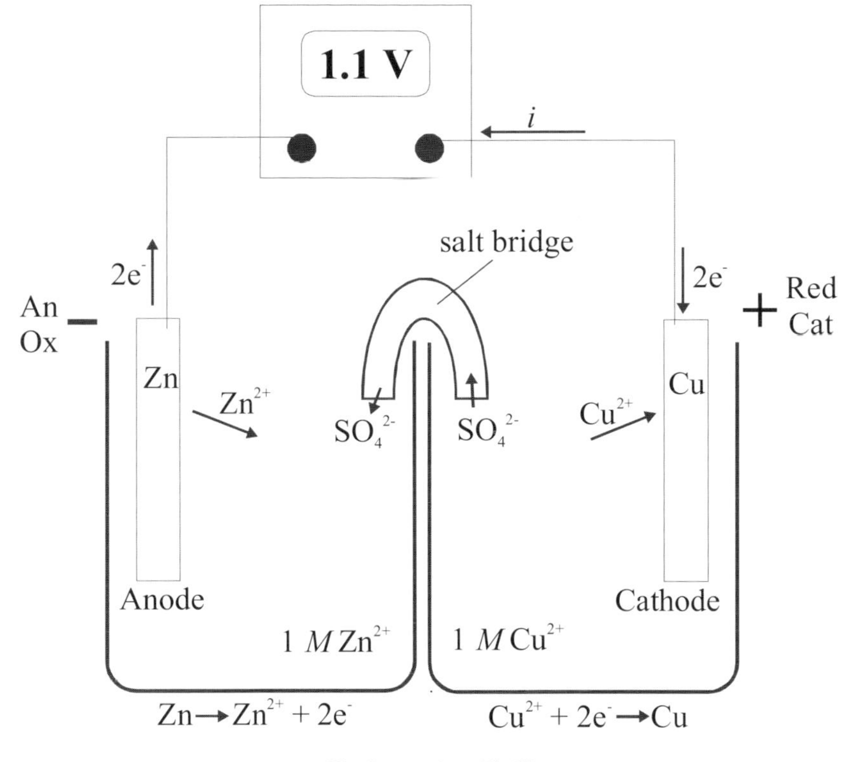

Galvanic Cell

A galvanic cell has two **electrodes**: the **anode** and the **cathode**. The oxidation half reaction takes place at the anode, and the reduction half reaction takes place at the cathode. As always, the current flows in the opposite direction of the electrons.

The **cell potential** $\mathscr{E}$, also called the **electromotive force (emf)**, is simply the sum of the half reaction potentials. The cell potential for a galvanic cell is always positive; a galvanic cell always has energy to give away. The cell potential depends upon the half reactions, the concentrations of the reactants and products, and the temperature. Since electrons in the anode have higher potential energy than those in the cathode, electrons flow through the external circuit from the anode to the cathode.

Free Energy and Chemical Energy

A positive cell potential indicates a spontaneous reaction as shown by the following equation:

$$\Delta G = -nF\mathscr{E}_{max}$$

where n is the number of moles of electrons that are transferred in the redox reaction, and F is Faraday's constant, which is the charge on one mole of electrons (96,486 C mol^{-1}). This equation simply represents the product of the total charge nF and the voltage $\mathscr{E}$. Recall from Physics Lecture 7 that the product of charge and voltage equals work ($w = qV$). Since this is electrical work, it is not the result of a change in pressure or volume, so it is nonPV work. Recall from Chemistry Lecture 3 that ΔG is the maximum amount of nonPV work available from a reaction at constant temperature and pressure. A negative ΔG indicates that the work is being done by the system and not on the system.

But why is this the maximum work? Go back and look once more at the table of reduction potentials. Notice that it is titled "Standard reduction potentials". These are the reduction potentials for these reactions at standard state. Recall from Chemistry Lecture 3 that the standard state for the aqueous phase is a one molar concentration. Notice in the diagram of the galvanic cell that both sides of the cell are at a one molar concentration. The moment before the current begins to flow, the voltage is 1.1 V. As we can see by the half reactions, as soon as current begins to flow, the concentrations of both sides will change. This means that the potential for a galvanic cell will change. The cell above starts at 1.1 V and decreases immediately as the concentration of reactants decreases. Thus, all the charge is not transferred at 1.1 V, and less work is done than predicted by $\Delta G = -nF\mathscr{E}$.

When all the conditions are standard, we can write the equation above using the '^{o}' symbol as follows (the 'max' is part of the definition of ΔG and is assumed):

$$\Delta G^{o} = -nF\mathscr{E}^{o}$$

ΔG^{o} can be found in books, but what about non standard state conditions? There are an infinite number of possible combinations of concentrations of reactants and products and temperatures with which we could start the reaction. How can we predict the maximum available work from these combinations? In order to make predictions about reactions that do not occur at standard state, we must use the following equation, which relates ΔG with ΔG^{o}:

$$\Delta G = \Delta G^{o} + RT\ln(Q)$$

where Q is the reaction quotient discussed in Chemistry Lecture 2, and 'ln( )' is the natural logarithm. You may see this equation written using a base 10 logarithm as:

$$\Delta G = \Delta G^{o} + 2.3RT\log(Q)$$

This is based upon the crude approximation: $2.3\log(x) \approx \ln(x)$.

There is no need to confuse ΔG and ΔG^o. ΔG^o is a specific ΔG with specifically described parameters called standard conditions. Notice that if we use only one molar concentrations for Q, $Q = 1$, and $RT\ln(Q) = 0$, leaving us with $\Delta G = \Delta G^o$. This is what we would expect for a reaction at standard conditions. (Remember, standard conditions don't actually indicate a particular temperature; you can have standard conditions at any temperature. Standard conditions are usually assumed to be 298 K.)

Recall from Chemistry Lecture 3 that at equilibrium, there is no available free energy with which to do work; $\Delta G = 0$ by definition. Thus, if we have equilibrium conditions, we can plug in a value of 0 for ΔG, and rewrite "$\Delta G = \Delta G^o + RT\ln(Q)$" as:

$$\Delta G^o = -RT\ln(K)$$

In this equation, both K and ΔG^o vary with temperature. Whenever you specify a new temperature, you must look up a new ΔG^o for that temperature. Notice that since this equation uses the natural log of the equilibrium constant, a value of 1 for K will result in a value of 0 for ΔG^o. For the MCAT you should understand the relationship between K and ΔG^o.

if	$K = 1$	**then**	$\Delta G^o = 0$
if	$K > 1$	**then**	$\Delta G^o < 0$
if	$K < 1$	**then**	$\Delta G^o > 0$

Warning: this relationship does not say that if a reaction has an equilibrium constant that is greater than one, the reaction is spontaneous. That doesn't make any sense, since no reaction at equilibrium is spontaneous. It does say that if a reaction has an equilibrium constant that is greater than one, the reaction is spontaneous at standard state (molar concentrations of exactly 1 *M*) and the prescribed temperature.

This is pretty tricky stuff, but it's worth wracking your brains and killing some time on it now, rather than on the MCAT. Reread this section and make sure that you understand the relationship between K, Q, ΔG^o, ΔG, and T.

The galvanic cell diagrammed earlier has standard conditions. That's great for the instant that the concentrations are all one molar, but what about for the rest of the time? How can we find the potential when the concentrations aren't one molar? If we take the equation:

$$\Delta G = \Delta G^{\circ} + RT\ln(Q)$$

and substitute $-nF\mathscr{E}$ for ΔG, and $-nF\mathscr{E}^{\circ}$ for ΔG°, and then divide by $-nF$, we get:

$$\mathscr{E} = \mathscr{E}^{\circ} - \frac{RT}{nF}\ln(Q)$$

This is the *Nernst equation*. At 298 K, and in base 10 logarithm form, the Nernst equation is:

$$\mathscr{E} = \mathscr{E}^{\circ} - \frac{0.06}{n}\log(Q)$$

You do not have to know the Nernst equation for the MCAT. However, you should understand how the Nernst equation expresses the relationship between chemical concentrations and potential difference. For instance, the Nernst equation could be used to express the resting potential across the membrane of a neuron. Such a situation would be similar to a concentration cell as discussed in the next section.

Questions 96 through 100 are **NOT** based on a descriptive passage.

96. Which of the following statements about a galvanic cell is false?

A. If $\mathscr{E}^o = 0$, a reaction may still be spontaneous depending upon the chemical concentrations.
B. A galvanic cell with a positive potential can perform work.
C. Reduction takes place at the cathode.
D. A salt bridge balances the charge by allowing positive ions to move to the anode.

97. The values of all of the following are reversed when a reaction is reversed EXCEPT:

A. enthalpy
B. Gibbs energy
C. the rate constant
D. reaction potential

98. Which of the following is true for a reaction, if $\Delta G^o_{298} = 0$? (The 298 subscript indicates a temperature of 298 K.)

A. The reaction is at equilibrium.
B. At 298 K and 1 *M* concentrations of products and reactants the equilibrium constant equals one.
C. ΔG is also zero.
D. The reaction is spontaneous at temperatures greater than 298 K.

99. The following is a table of half reactions:

Half reaction	$\mathscr{E}^o$ **(V)**
$Ag^{2+} + e^- \rightarrow Ag^+$	1.99
$Fe^{3+} + e^- \rightarrow Fe^{2+}$	0.77
$Cu^{2+} + 2e^- \rightarrow Cu$	0.34
$2H^+ + 2e^- \rightarrow H_2$	0.00
$Fe^{2+} + 2e^- \rightarrow Fe$	–0.44
$Zn^{2+} + 2e^- \rightarrow Zn$	–0.76

The strongest reducing agent shown in the table is:

A. Zn
B. Zn^{2+}
C. Ag^+
D. Ag^{2+}

100. A positive cell potential indicates which of the following:

A. Both half reactions are spontaneous.
B. The reduction half reaction potential is greater than the oxidation half reaction potential.
C. The oxidation half reaction potential is greater than the reduction half reaction potential.
D. The cell is voltaic.

Answers to Questions 96-100

96. D is correct. Positive ions move across the salt bridge to the cathode. You can remember this because the salt bridge is used to balance the charges. Since negative electrons move to the cathode, positive ions must balance the charge by moving to the cathode.

97. C is correct. The forward and reverse reaction rates are only equal at equilibrium, and their rate constants are rarely equal.

98. B is correct. This question requires knowledge of the equation: $\Delta G^o = -RT \ln(K)$. This equation is a statement about the relationship between ΔG^o and K at a specific temperature. If $\Delta G^o = 0$, then $K = 1$. The standard state for an aqueous solution is 1 M concentrations.

99. A is correct. The strongest reducing agent is the one most easily oxidized; thus we must reverse the equations and the signs of the potentials.

100. D is correct. A positive potential indicates a galvanic cell (also called a voltaic cell).

More Cells

With the Nernst equation we can make a special type of galvanic cell called a **concentration cell**. A concentration cell is simply a galvanic cell with the a reduction half reaction taking place in one half cell and the reverse of that half reaction taking place in the other.

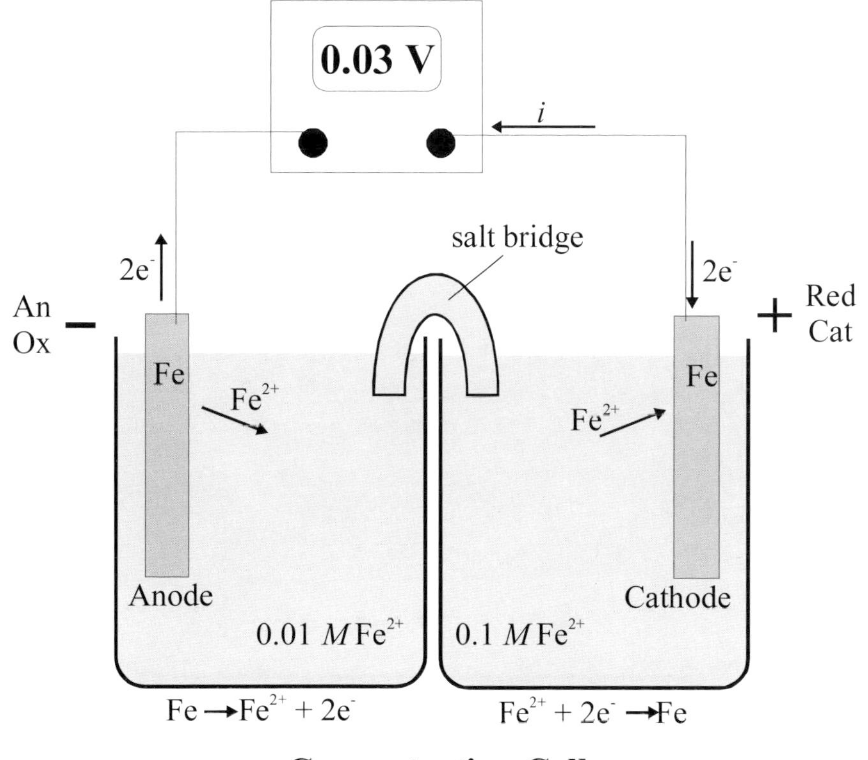

Concentration Cell

If the concentrations were equal on both sides, the cell potential would be zero. Of course, when we add the two half reactions we get: $\mathscr{E}^o = 0$. You can use the Nernst equation to find the potential for a concentration cell. (If you need the Nernst equation, the MCAT will give it to you.) It is much more likely that the MCAT will ask you a qualitative question like "In which direction will current flow in the concentration cell?" In this case, we must think about natures tendency for balance; nature wants to create the greatest entropy. The more concentrated side will try to become less concentrated, and electrons will flow accordingly.

To use the Nernst equation to find the potential of a concentration cell at 25 °C, we must realize that Fe^{2+} is both a product and a reactant. Thus, we simply substitute for Q the ratio of the Fe^{2+} concentrations on either side. For the case above we have:

$$\mathscr{E} = \mathscr{E}^o - \frac{0.06}{2}\log\left(\frac{0.01}{0.1}\right)$$

$n = 2$ because 2 electrons are used each time the reaction occurs, and $\mathscr{E}^o$ equals zero. Concentration cells tend to have small potentials.

If we hook up a power source across the resistance of a galvanic cell, and force the cell to run backwards, we have created another type of cell, the **electrolytic cell**. The electrolytic cell is a cell that is 'forced' to run by a power source. It does not have to have a negative voltage to be an electrolytic cell. It may be forced in forwards or backwards. Notice that since the electrolytic cell is run by an outside power source, using a salt bridge is possible but not necessary to running the cell.

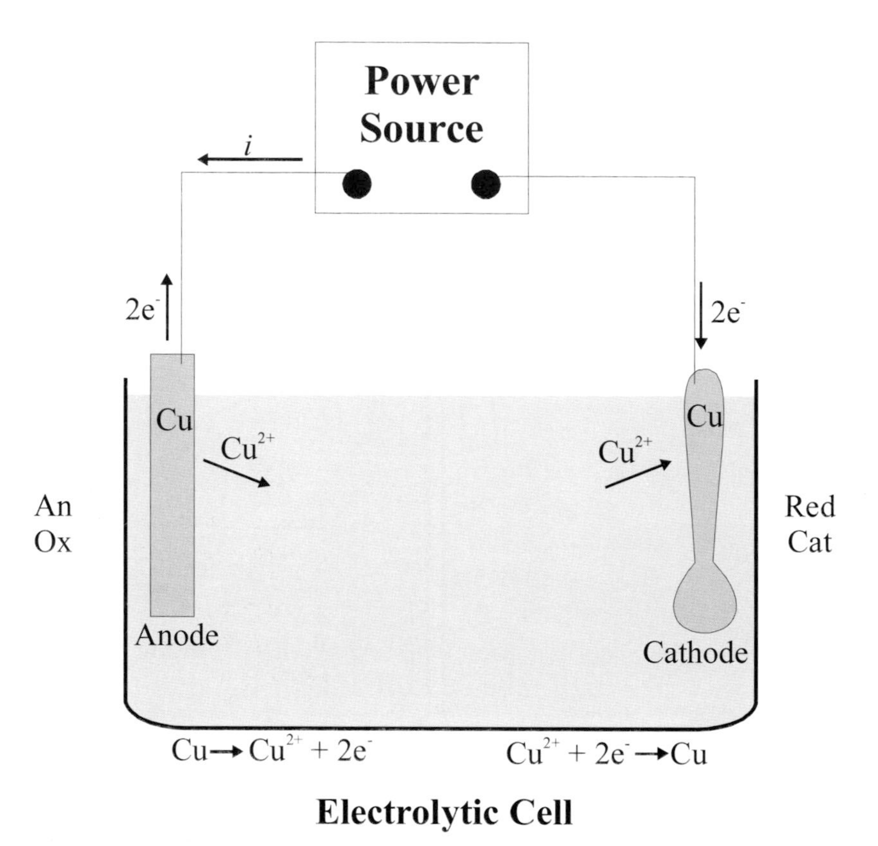

Electrolytic Cell

Electrolytic cells are used in industry for metal plating, and for purifying metals. For instance, pure sodium can be collected through electrolysis of sodium chloride solution in a *Downs cell.* The half reactions are as follows:

$$Na^+ + e^- \rightarrow Na \quad \mathscr{E}^o = -2.71\ V$$
$$2Cl^- \rightarrow 2e^- + Cl_2 \quad \mathscr{E}^o = -1.36\ V$$

Notice that this reaction will not run in aqueous solution because, from Table 7-3, we see that water has a less negative reduction potential than sodium. In fact, this indicates that solid sodium will oxidize spontaneously in water.

For cells, you should learn to diagram a galvanic cell by yourself. Once you can do that, the other cells can be created from the galvanic cell. Remember that galvanic cells have a positive cell potential; electrolytic cells may have either a positive or negative potential. Galvanic cells are spontaneous; electrolytic are forced by an outside power source.

For any and all cells, remember 'Red Cat, An Ox'. This translates to Reduction at the Cathode, and Oxidation at the Anode.

Questions 101 through 105 are **NOT** based on a descriptive passage.

101. A galvanic cell is prepared with solutions of Mg^{2+} and Al^{3+} ions separated by a salt bridge. A potentiometer reads the difference across the electrodes to be 1.05 Volts. The following standard reduction potentials at 25 °C apply:

Half Reaction	$\mathcal{E}^o$ (V)
$Al^{3+} + 3e^- \rightarrow Al$	–1.66
$Mg^{2+} + 2e^- \rightarrow Mg$	–2.37

Which of the following statements is true concerning the galvanic cell at 25 °C?

A. Magnesium is reduced at the cathode.
B. The concentrations of ions are 1*M*.
C. The reaction is spontaneous.
D. For every aluminum atom reduced, an equal number of magnesium atoms are oxidized.

102. Which of the following is true for an electrolytic cell?

A. Reduction takes place at the anode.
B. The reaction is spontaneous.
C. electrons flow to the cathode.
D. An electrolytic cell requires a salt bridge.

103. A concentration cell contains 0.5 *M* aqueous Ag^+ on one side and 0.1 *M* aqueous Ag^+ on the other. All of the following are true EXCEPT:

A. Electrons will move from the less concentrated side to the more concentrated side.
B. Electrons will move from the anode to the cathode.
C. As the cell potential moves toward zero, the concentrations of both sides will tend to even out.
D. $\Delta G > 0$

104. According to the Nernst equation:

$$\mathcal{E} = \mathcal{E}^o - \frac{0.06}{n}\log(\frac{[x]}{[y]})$$

if a concentration cell has a potential of 0.12 V, and a concentration of 0.1 *M* Ag^+ at the anode, what is the concentration of Ag^+ at the cathode?

A. 10^{-3} *M*
B. 10^{-1} *M*
C. 1 *M*
D. 10 *M*

105. A spoon is plated with silver in an electrolytic process where the half reaction at the cathode is:

$$Ag^+(aq) + e^- \rightarrow Ag(s) \quad \mathcal{E}^o = 0.8 \text{ V}$$

If the current *i* is held constant for *t* seconds, which of the following expressions gives the mass of silver deposited on the spoon? (*F* is Faraday's constant.)

A. 107.8 *itF*
B. 107.8 *it*/*F*
C. 107.8 *I*/*tF*
D. 107.8 *iF*/*t*

Answers to Questions 101-105

101. C is correct. Reactions in galvanic cells are always spontaneous. To find the reaction for this cell we must flip the more negative half reaction. Now we have a spontaneous cell.

$Al^{3+} + 3e^- \rightarrow Al$	-1.66
$Mg \rightarrow Mg^{2+} + 2e^-$	$\underline{2.37}$
	0.71

We also have to multiply the aluminum reaction by 2 and the magnesium reaction by 3. Notice, however, that we do not multiply their potentials.

$$2Al^{3+} + 3Mg \rightarrow 3Mg^{2+} + 2Al \qquad \mathscr{E}^o = 0.71\ V$$

Since the potential for this cell does not equal this, the conditions must not be standard.

102. C is correct. Reduction always takes place at the cathode in any cell. This means that the cathode gains electrons.

103. D is correct. A concentration cell is a special type of galvanic cell. It is always spontaneous. The concentrations in the cell even out at equilibrium.

104. D is correct. In this cell the cathode has the greater concentration because electrons flow toward it to reduce the number of cations. Also in a concentration cell $\mathscr{E}^o = 0$, since the reduction half reaction is simply the reverse of the oxidation half reaction. $n = 1$ because only one electron is transferred in each reaction. x/y must be a fraction so that the log will be negative and $\mathscr{E}$ will be positive. Thus we have:

$$\mathscr{E} = -(0.06/1)\log(0.1/y) = 0.12$$

$$y = 10 \text{ so that } x/y = 10^{-2}.$$

105. B is correct. Use units to solve the problem. We want to go from current to grams. Current is C/s. F is coulombs per mol of electrons. For every mol of electrons there is one mol of silver. The molecular weight of silver is 107.8 g/mol

C/s x s = C => C x mol/C = mol => mol x grams/mol = grams

so:

i x t x $1/F$ x 107.8 = grams

STOP!

Do not look at these exams until class.

30-minute In-class Exam for Lecture 1

Passage I (Questions 1-7)

There are five types of interactions within and between molecules. Intramolecular interactions include covalent and ionic bonds. Intermolecular interactions include Van der Waals's forces, dipole-dipole, and hydrogen bonds. Table 1 lists the typical energies for these interactions.

Interaction	Typical Energy kJ mol^{-1}
Van der Waals's	0.1 – 5
Dipole-dipole	5 – 20
Hydrogen bond	5 – 50
Ionic bond	400 – 500
Covalent bond	150 – 900

Table 1. Energies of interactions

The boiling point of a substance increases with the strength of its intermolecular bonds. Figure 1 shows the boiling points of hydrides for some main-group elements and of the noble gases.

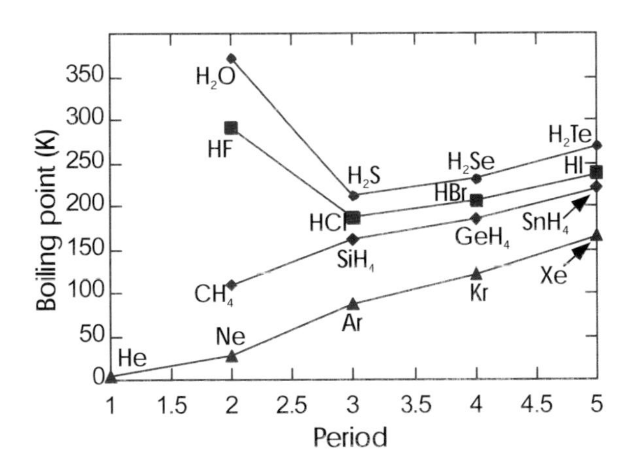

Figure 1. Boiling points of some main-group hydrides and the noble gases.

1. Why does the boiling point of H_2O and HF deviate from the trend in Figure 1?

   A. F and O both occur in the second period.
   B. The size of H_2O and HF are small relative to the other molecules
   C. H_2O and HF are less polarizable.
   D. H_2O and HF can hydrogen bond.

2. What type of bonding holds together the compound $MgCl_2$?

   A. covalent
   B. ionic
   C. hydrogen
   D. Van der Waals's

3. Why do the boiling points of the noble gases increase as the period increases?

   A. The bonds are stronger because larger atoms are more polarizable as period increases.
   B. The bonds are weaker because larger atoms are more polarizable as period increases.
   C. The bonds are stronger because larger atoms are less polarizable as period increases.
   D. The bonds are weaker because larger atoms are less polarizable as period increases.

4. The atomic radius of Ne is:

   A. greater than Ar
   B. less than Ar
   C. the same as Ar
   D. cannot be determined

5. Why are boiling points a better indication of intermolecular bonding than melting points?

   A. Vaporization requires more energy than melting.
   B. Vaporization requires less energy than melting.
   C. Transition from solid to liquid involves other factors such as crystalline lattice structures.
   D. Vaporization is easier to measure.

6. What type of intermolecular bonding occurs in gaseous CH_4?

   A. covalent
   B. ionic
   C. hydrogen
   D. Van der Waals's

7. Why is a dipole-dipole interaction stronger than a Van der Waals's interaction?

   A. Dipole-dipole is an electrostatic interaction.
   B. Van der Waals's interactions rely on temporarily induced dipoles.
   C. Van der Waals's interactions require a large surface area.
   D. Dipole-dipole interactions only occur with ionically bonded compounds.

GO ON TO THE NEXT PAGE.

Passage II (Questions 8-15)

Figure 1 shows atomic radius as a function of atomic number for the first three periods of the periodic table. Within a period the atomic radius decreases as the atomic number increases, but the atomic radius increases as the period increases.

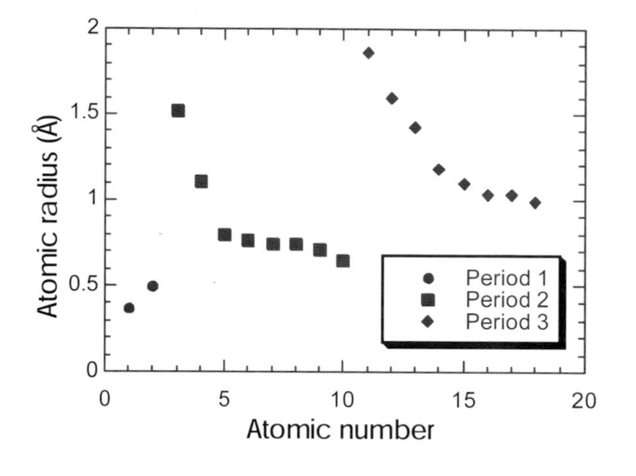

Figure 1. Atomic radius as a function of atomic number

Electronegativity of an atom also follows a trend in the periodic table. Electronegativity for any element (X) is based upon the difference (Δ) between the actual bond energy of a bond between element X and hydrogen and the expected bond energy of the same bond:

$$\Delta = (H{-}X)_{\text{actual bond energy}} - (H{-}X)_{\text{expected bond energy}}$$

where the expected bond energy is given by:

$$H{-}X_{\text{exp bond energy}} = \frac{H{-}H_{\text{bond energy}} + X{-}X_{\text{bond energy}}}{2}$$

Pauling electronegativity values are assigned to each element based upon its Δ value with respect to fluorine. Flourine is arbitrarily given a value of 4.0. The electronegativity of elements may be used to predict the type of bonding found in a molecule. Large differences in electronegativities of atoms in a bond result in an ionic bond.

8. If Se has an atomic radius of 1.16 Å, what is the predicted atomic radius of As?

A. 1.05 Å
B. 1.15 Å
C. 1.25 Å
D. 1.97 Å

9. Which of the following elements is the most chemically similar to Na?

A. H
B. Mg
C. C
D. Cs

10. Which element has the largest atomic radius?

A. Li
B. Ne
C. Rb
D. Br

11. Why is the electronegativity scale adjusted to fluorine?

A. the researcher who discovered electronegativity was working with fluorine.
B. fluorine has the smallest atomic radius
C. fluorine has the smallest electronegativity
D. fluorine has the greatest electronegativity

12. How does electron affinity change with atomic number?

A. Electron affinity becomes more exothermic as atomic number increases in a period and a group.
B. Electron affinity becomes less exothermic as atomic number increases in a period and a group.
C. Electron affinity becomes more exothermic as atomic number increases in a period and less exothermic as atomic number increases in a group.
D. Electron affinity becomes less exothermic as atomic number increases in a period and more exothermic as atomic number increases in a group.

13. Why does the atomic radius follow the trends observed in Figure 1?

A. As the atomic number increases, the nuclear charge increases.
B. As the atomic number increases, the nuclear charge decreases.
C. As the atomic number increases, the nuclear charge increases and as the period increases the number of electron shells increases.
D. As the atomic number increases, the nuclear charge increases and as the period increases the number of electron shells decreases.

GO ON TO THE NEXT PAGE.

14. Hydrogen has a Pauling electronegativity of 2.1. What is the value of Δ for hydrogen?

A. 0
B. 1.0
C. 2.1
D. 4.0

15. What type of intramolecular bonding is found in a CO molecule?

A. covalent
B. ionic
C. hydrogen
D. Van der Waals's

Passage III (Questions 16-21)

The empirical formula of a hydrocarbon can be determined using an instrument similar to the one shown in Figure 1. A sample hydrocarbon is combusted. The absorption chambers absorb all the water and carbon dioxide from the reaction. $CaCl_2$ can absorb both water and CO_2. The masses of the chambers before and after the reaction are compared to find the moles of carbon and hydrogen in the sample.

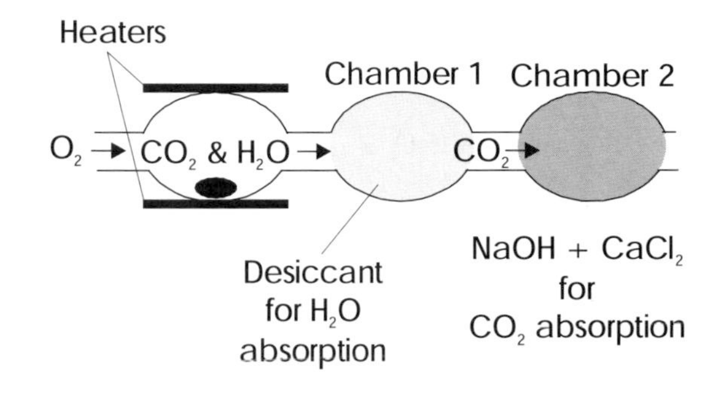

Figure 1. A combustion train

For example: propane can be combusted in the apparatus as follows:

$$C_3H_8 + 5O_2 \rightarrow 4H_2O + 3CO_2$$

In an experiment using the combustion train, a gaseous fuel used in welding (containing only C and H) is reacted with O_2. The mass of the absorbers in Chamber 1 increases by 0.9 grams and the mass of the absorbers in chamber 2 increases by 4.4 grams.

The density of the welding gas is 1.1 g L^{-1} at 25 °C and atmospheric pressure. At the same conditions, O_2 has a density of 1.3 g L^{-1}.

16. What is the empirical formula of the welding gas?

A. CHO
B. CH
C. C_2H_2
D. C_3H_8

17. What is the molecular weight of the gas?

A. 13 g/mol
B. 26 g/mol
C. 32 g/mol
D. 60 g/mol

GO ON TO THE NEXT PAGE.

18. A compound has an empirical formula of CH_2O. Using osmotic pressure, the molecular weight is determined to be 120 g/mol. What is the molecular formula for this compound?

A. CH_2O
B. $C_4H_4O_3$
C. $C_3H_6O_3$
D. $C_4H_8O_4$

19. If 1 mole of C_3H_8 is reacted with 2.5 moles of O_2, how many moles of H_2O will be produced?

A. 1 mole of H_2O
B. 2 moles of H_2O
C. 3 moles of H_2O
D. 4 moles of H_2O

20. What would happen if the order of the chambers in the combustion train were reversed?

A. The amount of CO_2 calculated would be higher than the actual amount produced.
B. The amount of CO_2 calculated would be lower than the actual amount produced.
C. The amount of H_2O calculated would be higher than the actual amount produced.
D. Nothing, the experiment would still give the same results.

21. Why is it necessary to react the O_2 in excess when using a combustion train?

A. In addition to the combustion reaction, the O_2 is used as a carrier gas.
B. In addition to the combustion reaction, the O_2 is used as a source of energy to propel the non-spontaneous reaction.
C. O_2 needs to be the limiting reagent in order for the calculations to be correct.
D. The sample needs to be the limiting reagent in order for the calculations to be correct.

Questions 22 through 23 are **NOT** based on a descriptive passage.

22. What is the electron configuration of a chloride ion?

A. [Ne] $3s^2\ 3p^5$
B. [Ne] $3s^2\ 3p^6$
C. [Ne] $3s^2\ 3d^{10}\ 3p^5$
D. [Ar] $3s^2\ 3p^6$

23. According to the Heisenberg uncertainty principle, which of the following pairs of properties of an electron cannot be known with certainty at the same time?

A. charge and velocity
B. spin and subshell
C. average radius and energy level
D. momentum and position

STOP. IF YOU FINISH BEFORE TIME IS CALLED, CHECK YOUR WORK. YOU MAY GO BACK TO ANY QUESTION IN THIS TEST BOOKLET.

STOP.

30-minute In-class Exam for Lecture 2

Passage I (Questions 24-29)

Over the years, many attempts have been made to find an equation which represents the behavior of non-ideal gases. Although none of the equations are completely accurate, they do allow an investigation of some of the macroscopic properties of real gases. The most commonly used of these is the *van der Waals equation:*

$$\left(P + a\frac{n^2}{V^2}\right)(V - nb) = nRT$$

where P is the absolute pressure, n is the number of moles, V is the volume, T is the absolute temperature, R is the gas constant (0.08206 L atm/mol K), and a and b are constants determined experimentally for each gas studied. Table 1 gives the values of a and b for some common gases:

Gas	a (atm L^2/mol)	b (L/mol)
Ar	1.4	0.032
HCl	3.7	0.041
Cl_2	6.4	0.054
H_2	0.25	0.027
NH_3	4.3	0.037
O_2	1.4	0.032

Table 1 van der Waals Constants for Various gases

To help quantify the deviation of a real gas from ideality, a *compression factor* Z has been defined by Z = PV/nRT. Figure 1 shows how the compression factor for ammonia depends on pressure at several different temperatures.

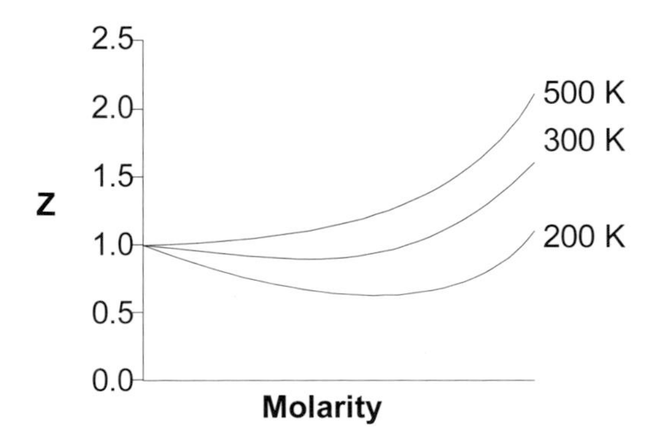

Figure 1 Compression factors for ammonia

24. For an ideal gas, which of the following is most likely the correct graph?

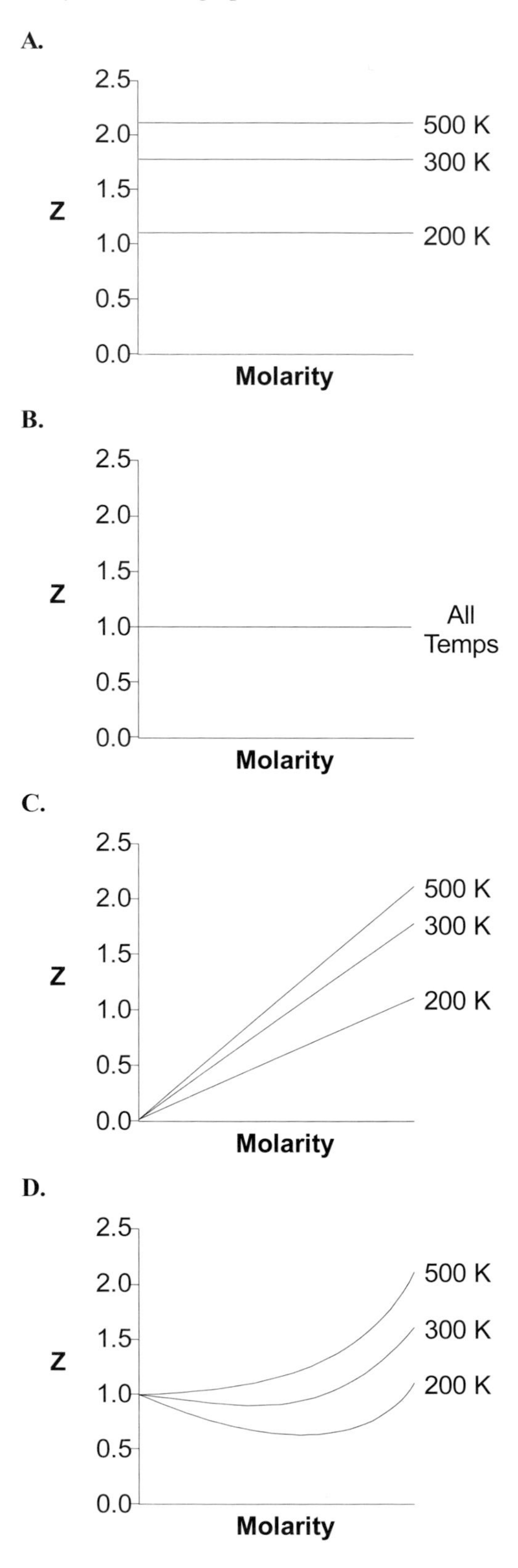

GO ON TO THE NEXT PAGE.

25. For an ideal gas, what can be said about the constants a and b?

A. They are both zero.
B. They are both positive and equal to each other.
C. They depend on the temperature of the gas.
D. They must be determined experimentally.

26. Based on the information in the passage, under which of the following conditions does ammonia behave most ideally?

A. Low temperatures and low pressures
B. Low temperatures and high pressures
C. High temperatures and low pressures
D. High temperatures and high pressures

27. Which of the following statements is NOT true for an ideal gas?

A. The average kinetic energy of the molecules depends only on the temperature of the gas.
B. At constant volume in a sealed container, the pressure of the gas is directly proportional to its temperature.
C. At constant temperature in a sealed container, the volume of the gas is directly proportional to its pressure.
D. The intermolecular attractions between the gas molecules are negligible.

28. Which of the following demonstrates nonideal behavior of a gas?

A. Some of the molecules move more rapidly than others.
B. Condensation occurs at low temperatures.
C. The gas exerts a force on the walls of its container.
D. The average speed of the molecules in the gas is proportional to the square root of the absolute temperature.

29. Why must absolute temperature be used in the van der Waals equation?

A. Because the van der Waals equation is a nonrelativistic equation.
B. Because it is impossible to have a negative absolute temperature.
C. Because ratios of temperatures on other scales, such as the Celsius scale, are meaningless.
D. Because international convention requires it.

Passage II (Questions 30-37)

In 1889, Svante Arrhenius proposed that the rate constant for a given reaction is given by the formula:

$$k = Ae^{-\frac{E_a}{RT}}$$

where E_a is the *activation energy* for the reaction, R is the gas constant (8.314 J/mol K), T is the absolute temperature, and A is a factor, which depends on factors such as molecular size. *Catalysts* change the reaction pathway, which may result in a change in E_a, A, or both.

In *heterogeneous catalysis*, the catalyst is in a different phase from the reactants and products. For example, a solid may catalyze a fluid-phase reaction. Such a catalysis involves the following steps:

1) A reactant molecule diffuses through the liquid to the surface of the catalyst.

2) The reactant molecule bonds to the catalyst (*adsorption*).

3) Adsorbed molecules bond with each other or with a molecule which collides with the adsorbed molecules.

4) The product leaves the catalyst.

In *homogeneous catalysis*, the catalyst is in the same phase as the reactants and products. Acids often act by this mechanism.

30. Which of the following is true?

A. For a given reaction, the rate constant is affected only by the temperature.
B. Bases may act as homogeneous catalysts.
C. Catalysts increase the proportion of products present in equilibrium.
D. The value of R may depend on temperature.

GO ON TO THE NEXT PAGE.

31. Consider the following mechanism:

$Cl_2 \rightarrow 2Cl$
$Cl + CO \rightarrow COCl$
$COCl + Cl_2 \rightarrow COCl_2 + Cl$

In this mechanism, what is the catalyst?

A. Cl_2
B. Cl
C. COCl
D. No catalyst is shown in this mechanism

32. As the temperature of a reaction increases, which of the following always occurs?

A. The rate constant increases.
B. The rate constant decreases.
C. The activation energy increases.
D. The activation energy decreases.

33. Consider a reversible reaction. If the activation energy for the forward reaction is lowered by a catalyst, what can be said about the activation energy for the reverse reaction?

A. It is also lowered.
B. It is raised.
C. It is unaffected by the catalyst.
D. The effect of the catalyst on the reverse reaction cannot be predicted without more information.

34. Suppose a reaction is acid-catalyzed by a solution of pH 3.0. What can be said about the pH of the resulting solution?

A. It will be greater than 3.0 because the acid is consumed.
B. It will be equal to 3.0 because the acid is regenerated.
C. It will be equal to 3.0 because catalysts have no effect on equilibrium.
D. It cannot be predicted without information on the acidity of the reactants and products.

35. The rate of a reaction may depend on which of the following?

I. Concentrations of the reactants
II. Concentration of a catalyst
III. Surface area of a heterogeneous catalyst
IV. Temperature

A. I only
B. IV only
C. I and IV only
D. I, II, III, and IV

36. H_2 can be added to ethylene in the presence of a heterogeneous catalyst such as solid platinum. What might account for the initial attraction between the hydrogen molecules and the solid platinum?

A. hydrogen bonding
B. metallic bonding
C. van der Waals attraction
D. the plasma continuum effect

37. If the solid line in the graph below represents the reaction profile for an uncatalyzed reaction, which line might represent the reaction profile for the catalyzed reaction?

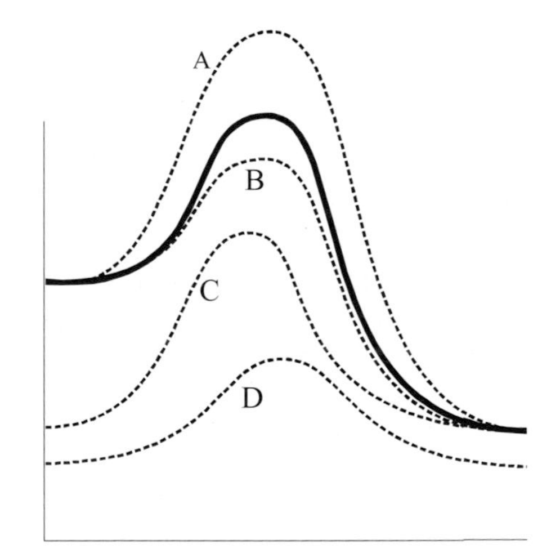

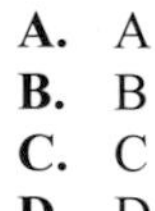

A. A
B. B
C. C
D. D

GO ON TO THE NEXT PAGE.

Passage III (Questions 38-44)

Peroxydisulfate (persulfate) ion reacts with the iodide anion according to Reaction 1.

$$S_2O_8^{2-}\,(aq) + 3I^-\,(aq) \rightleftharpoons 2SO_4^{2-}\,(aq) + I_3^-\,(aq)$$

Reaction 1

The amount of I_3^- formed can be determined by adding a known amount of $S_2O_3^{2-}$ and allowing it to react according to Reaction 2.

$$2S_2O_3^{2-}\,(aq) + I_3^-\,(aq) \rightleftharpoons S_4O_6^{2-}\,(aq) + 3I^-\,(aq)$$

Reaction 2

If starch is also added, any excess I_3^- will react to form a blue-black I_2 complex. The formation of this complex indicates the completion of Reaction 2. The rate of Reaction 1 can be determined by the following equation where t is the elapsed time from the addition of the last component to the formation of the blue-black starch, I_2 complex.

$$rate = \frac{\frac{1}{2}[S_2O_3^{2-}]}{t}$$

Equation 1

38. Why can Equation 1 be used to measure the rate of Reaction 1?

- **A.** Reaction 2 must be much faster than Reaction 1, thus the rate in Equation 1 is the rate of formation of I_3^-.
- **B.** Reaction 2 must be much slower than Reaction 1, thus the rate in Equation 1 is the rate of formation of I_3^-.
- **C.** Reactions 1 and 2 must occur at the same rate, thus the rate in Equation 1 is the rate of formation of I_3^-.
- **D.** Equation 1 can be derived directly from the rate laws of Reactions 1 and 2.

39. What would happen to the time and the rate in Equation 1, if the temperature were reduced?

- **A.** Time would increase and rate would decrease.
- **B.** Time would decrease and rate would increase
- **C.** Time would increase and rate would remain unchanged.
- **D.** Time would remain the same and rate would increase.

40. The following table gives the relative concentrations and rates found using the method described in the passage. What is the rate law for Reaction 1?

	1	**2**	**3**
$[I^-]$, (*M*)	0.060	0.030	0.030
$[S_2O_8^{2-}]$, (*M*)	0.030	0.030	0.015
Rate, (*M*/sec)	6.0 x 10^{-6}	3.0 x10^{-6}	1.5 x 10^{-6}

- **A.** $k[I^-]^2\,[S_2O_8^{2-}]^2$
- **B.** $k[I^-]^{1/2}\,[S_2O_8^{2-}]^2$
- **C.** $k[I^-]^3\,[S_2O_8^{2-}]$
- **D.** $k[I^-]\,[S_2O_8^{2-}]$

41. The rate expression for the reaction of H_2 with Br_2 is:

rate = $k[H_2][Br_2]$.

- **A.** The rate is first order with respect to H_2, and first order overall.
- **B.** The rate is first order with respect to H_2, and second order overall.
- **C.** The rate is second order with respect to H_2, and first order overall.
- **D.** The rate is second order with respect to H_2, and second order overall.

GO ON TO THE NEXT PAGE.

42. A student is performing the kinetic study described in the passage and forgets to add starch. What will be the result of the experiment?

A. The rates of both reactions as measured by the student will increase because the starch slows the reactions.
B. The rate of Reaction 1 as measured by the student will decrease because starch speeds up the reaction.
C. The rate of Reaction 1 as measured by the student will stay the same because starch has no effect on the rate.
D. The rate of Reaction 1 as measured by the student will not be able to be determined by the method described in the passage.

43. What would happen to the time and the rate in Equation 1, if more $S_2O_3^{2-}$ were added, and all other conditions remained the same?

A. Time would increase and rate would decrease.
B. Time would decrease and rate would increase.
C. Time would increase and rate would remain unchanged.
D. Time would remain the same and rate would increase.

44. What would happen to the time and the rate in Equation 1, if a catalyst is added to Reaction 1, and all other conditions remain the same?

A. Time would increase and rate would decrease.
B. Time would decrease and rate would increase.
C. Time would increase and rate would remain unchanged.
D. Time would remain the same and rate would increase.

Questions 45 through 46 are **NOT** based on a descriptive passage.

45. Which of the following are true concerning any reaction at equilibrium?

I. The concentration of products is equal to the concentration of reactants.
II. The rate of change in the concentration of the products is equal to the rate of change in the concentration of reactants.
III. The rate constant of the forward reaction is equal to the rate constant of the reverse reaction.

A. II only
B. I and II only
C. II and III only
D. I, II, and III

46. Equal concentrations of hydrogen and oxygen gas are placed on side 1 of the container shown below. Side 2 contains a vacuum. A small pin hole exists in the barrier separating side 1 and side 2. Which of the following statements is true?

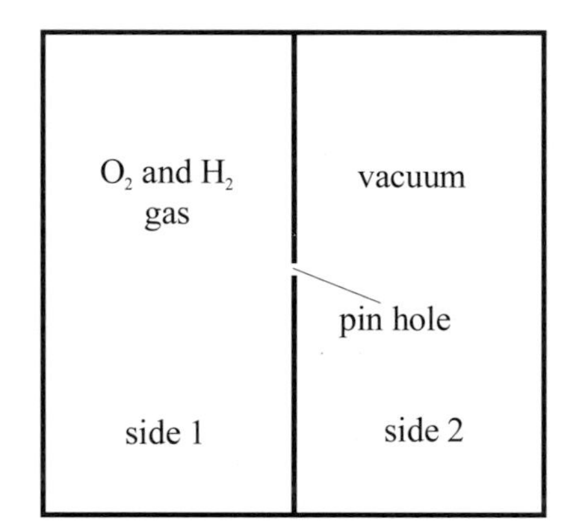

A. The partial pressure of oxygen on side 1 will increase.
B. The partial pressure of hydrogen on side 1 will increase.
C. The mole fraction of oxygen on side 1 will increase.
D. The mole fraction of hydrogen on side 1 will increase.

STOP. IF YOU FINISH BEFORE TIME IS CALLED, CHECK YOUR WORK. YOU MAY GO BACK TO ANY QUESTION IN THIS TEST BOOKLET.

STOP.

30-minute In-class Exam for Lecture 3

Passage I (Questions 48-52)

Nickel is purified by the Mond process, which relies on the equilibrium:

$$\text{Ni}(s) + 4\text{CO}(g) \rightleftharpoons \text{Ni(CO)}_4(g)$$

$$\Delta H^\circ = -160.8 \text{ kJ}, \Delta S^\circ = -409.5 \text{ JK}^{-1} \text{ at } 25\ ^\circ\text{C}$$

Reaction 1 The Mond process

Two chemists analyze the equilibrium.

Chemist A

Chemist A argues that Reaction 1 will be spontaneous in the forward direction because the product is more stable than the reactants. Furthermore, if the temperature is raised, the reaction will run in reverse because it is an exothermic reaction.

Chemist B

Chemist B argues that Reaction 1 will be spontaneous in the reverse direction because the entropy is higher for the reactants than for the products. Furthermore, if the temperature is raised, the spontaneity of the reverse reaction will increase.

47. The ΔH°_f for Ni(s) is:

A. 0 kJ/mol
B. -160.8 kJ/mol
C. 160.8 kJ/mol
D. 409.5 kJ/mol

48. Which of the following is a logical conclusion of Chemist A's argument?

A. The enthalpy change of a reaction is an indicator of the relative stability of the reactants and products.
B. The entropy change of a reaction is an indicator of the relative stability of the reactants and products.
C. At higher temperatures, the reactants of Reaction 1 will be more stable than the products.
D. Reactions do not necessarily need to increase the entropy of the universe in order to be spontaneous.

49. What are the most efficient conditions for purifying nickel when using the Mond process?

A. low pressure and low temperature
B. low pressure and high temperature
C. high pressure and low temperature
D. high pressure and high temperature

50. Which of the following explains the error in Chemist B's argument?

A. Higher temperatures favor a reaction that decreases entropy.
B. The entropy change of a reaction is the entropy change of the universe.
C. A positive entropy of reaction does not necessarily indicate an entropy increase for the universe.
D. The reaction will only run in the reverse direction until entropy is maximized for the reaction system.

51. What is the change in Gibbs free energy for Reaction 1 at standard state and 25 °C?

A. –38 kJ
B. 38 kJ
C. –150.5 kJ
D. 150.5 kJ

52. Consider the reaction below:

$$\text{C(s)} + \text{O}_2\text{ (g)} \rightarrow \text{CO}_2\text{ (g)}$$

$$\Delta H^\circ \text{ is } -393.51 \text{ kJ and } \Delta S^\circ \text{ is } 2.86 \text{ JK}^{-1} \text{ at } 25\ ^\circ\text{C}$$

If solid carbon is exposed to 1atm of oxygen and 1 atm of carbon dioxide gas at room temperature, will carbon dioxide gas form spontaneously?

A. No, because the enthalpy of formation for CO_2 is negative.
B. No, because the change in Gibbs energy is negative.
C. Yes, because the enthalpy of formation for CO_2 is negative.
D. Yes, because the change in Gibbs energy is negative.

GO ON TO THE NEXT PAGE.

Passage II (Questions 53-58)

A heat engine converts heat energy to work via a cyclical process which necessarily results in some of the heat energy being transferred from a higher temperature heat reservoir to a lower temperature heat reservoir. A heat engine obeys the *First Law of Thermodynamics*. The efficiency e of a heat engine is the fraction of heat energy input converted to useful work and is given by:

$$e = W/Q_h$$

where W is the work done by the heat engine on the surroundings and Q_h is the heat removed from the higher temperature reservoir. Work on the surroundings can also be represented by:

$$W = Q_h - Q_c$$

where Q_c is the heat energy expelled into the cold reservoir.

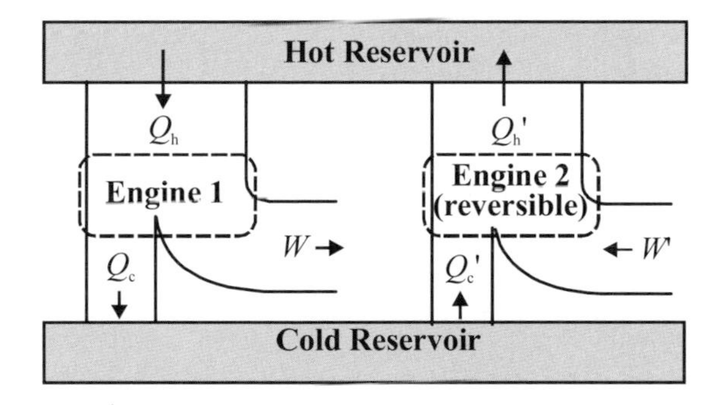

Figure 1 Two heat engines operating between the same heat reservoirs. Engine 2 is a Carnot engine operating in reverse.

The *Carnot theorem* states: No engine working between two heat reservoirs can be more efficient than a reversible engine working between those same reservoirs. Such a reversible engine is called a Carnot engine. This theorem can be proven with the *Second Law of Thermodynamics*, which states that for any process other than a reversible process the entropy change of the universe is greater than zero, and for a reversible process the entropy change of the universe is zero. Thus a Carnot engine obeys the following equation:

$$\Delta S = \frac{Q_c}{T_c} - \frac{Q_h}{T_h} = 0$$

Figure 1 shows a schematic representation of two heat engines working between the same two reservoirs. Engine 1 is absorbing heat energy from the hot reservoir and doing work while emitting heat energy into the cold reservoir. Engine 2 is a Carnot engine and is being run backwards removing heat energy from the cold reservoir and rejecting heat energy into the hot reservoir.

53. If both Engines 1 and 2 are operating at the same time as shown in Figure 1, and the rate of heat energy being removed from each reservoir is equal to the rate of heat energy being added then:

A. $W' > W$.
B. $W' < W$.
C. Engine 1 is a Carnot engine.
D. The efficiency of Engine 2 is greater than the efficiency of Engine 1.

54. Which of the following is true of the engines in Figure 1 if Engine 1 is not a Carnot engine and the work from Engine 1 is used to run Engine 2?

A. $Q_c < Q_c'$ and $Q_h < Q_h'$.
B. $Q_c > Q_c'$ and $Q_h < Q_h'$.
C. $Q_c > Q_c'$ and $Q_h > Q_h'$.
D. The work from Engine 1 cannot be used to run Engine 2.

55. Assume Engine 2 is running in the opposite direction as shown in Figure 1. Which of the following changes to Engine 2 will increase its efficiency as a heat engine?

A. increasing Q_h'
B. decreasing Q_c'
C. decreasing the temperature difference between the reservoirs
D. increasing the temperature difference between the reservoirs

56. If 50 J of heat energy are available to add to 10 liters of an ideal gas initially at 300 K, what is the maximum possible work attainable from the 50 J of heat energy? (Assume ideal conditions.)

A. 0 J
B. 30 J
C. 50 J
D. 500 J

GO ON TO THE NEXT PAGE.

57. What is the minimum power required for a heat engine to lift a 80 kg mass 5 m in 20 s if it releases 1000 J of heat energy from its exhaust each second?

A. 200 W
B. 500 W
C. 1200 W
D. 3000 W

58. A certain Carnot engine requires 18 kg of water in the form of steam as its working substance. When $5\text{x}10^5$ J of heat energy are added at a constant temperature of 400 K the gas expands to 4 m^3. What is the approximate pressure of the gas after the initial expansion? (The ideal gas constant is $R = 8.314$ J/K mol)

A. $8.3\text{x}10^5$ Pa
B. $8.3\text{x}10^7$ Pa
C. $1.3\text{x}10^6$ Pa
D. $1.3\text{x}10^8$ Pa

Passage III (Questions 59-65)

As shown in Figure 1, Reaction 1 is thermodynamically favored because the products are at a lower energy state than the reactants. However, at low temperatures this reaction will be too slow to be observed because the reactant molecules do not have enough energy to form the activated complex.

$$NO_2 + CO \rightarrow NO + CO_2$$

Reaction 1

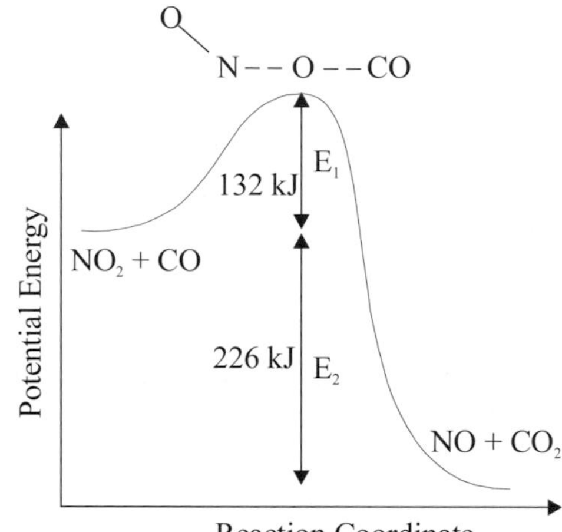

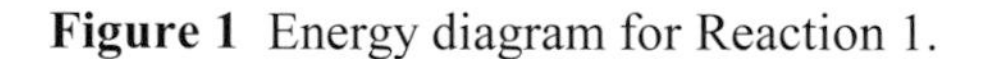
Figure 1 Energy diagram for Reaction 1.

As shown in Figure 2, Reaction 2 is not thermodynamically favored, but it has a smaller reaction barrier. The reactants require less energy to form the activated complex.

$$H_2 + Br \rightarrow H + HBr$$

Reaction 2

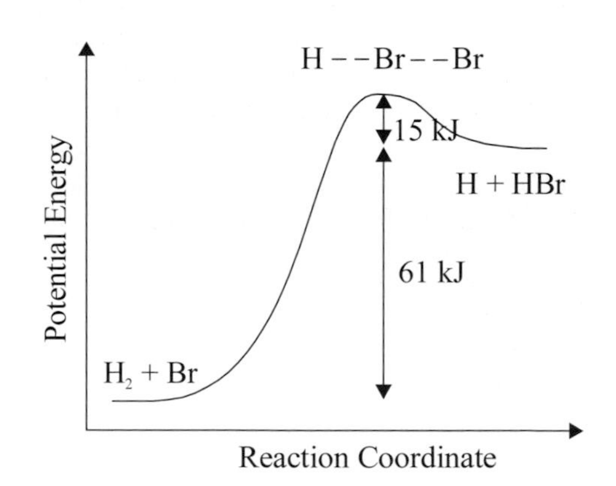

Figure 2 Energy diagram for Reaction 2

GO ON TO THE NEXT PAGE.

59. What does the symbol H- -Br- -Br represent in Figure 2?

A. an intermediate
B. a transition state
C. a reactant of the second step of the reaction
D. a product of the first step of the reaction

60. What is E_a for Reaction 2 as written?

A. 15 kJ/mole
B. 61 kJ/mole
C. 76 kJ/mole
D. 132 kJ/mole

61. Which reaction is most kinetically favored?

A. $NO_2 + CO \rightarrow NO + CO_2$
B. $NO + CO_2 \rightarrow NO_2 + CO$
C. $H_2 + Br \rightarrow H + HBr$
D. $H + HBr \rightarrow H_2 + Br$

62. Which reaction is most thermodynamically favored?

A. $NO_2 + CO \rightarrow NO + CO_2$
B. $NO + CO_2 \rightarrow NO_2 + CO$
C. $H_2 + Br \rightarrow H + HBr$
D. $H + HBr \rightarrow H_2 + Br$

63. If a catalyst were added to Reaction 1, what would happen?

A. E_1 would be less than 132 kJ but E_2 would remain unchanged.
B. E_1 would be less than 132 kJ and E_2 would be less than 226 kJ
C. E_2 would be less than 226 but E_1 would remain unchanged.
D. A catalyst doesn't affect the thermodynamic properties of the reactants and products. E1 and E2 would remain unchanged but the reaction rate would increase.

64. What is ΔE for Reaction 2 as written?

A. 15 kJ/mole
B. 61 kJ/mole
C. 76 kJ/mole
D. –76 kJ/mole

65. For the reaction,

+ HBr ⇌ (Br) **1** + Br **2**

the ratio of products changes with temperature. At -80 °C, 80% of 1 and 20% of 2 form. At 40 °C, 15% of 1 and 85% of 2 form. Assuming the relative stability of products vs. reactants does not change significantly with the change in temperature, which product is the kinetically favored one?

A. 1
B. 2
C. both 1 and 2
D. neither 1 nor 2

GO ON TO THE NEXT PAGE.

Questions 66 through 69 are **NOT** based on a descriptive passage.

66. An iron skillet is laid on a hot stove. After a few minutes the handle gets hot. The method of heat transfer described is:

A. convection.
B. conduction.
C. radiation.
D. translation.

67. A man straightens up his room. His action does not violate the second law of thermodynamics because:

A. the entropy of his room increased.
B. energy of the universe was conserved.
C. the entropy increase by the breakdown of nutrients in his body is greater than the entropy decrease by the straightening of his room.
D. His action does violate the second law of thermodynamics.

68. A metal rod is in thermal contact with two heat reservoirs both at constant temperature, one at 100 K and the other at 200 K. The rod conducts 1000 J of heat from the warmer to the colder reservoir. If no energy is exchanged with the surroundings, what is the total change of entropy?

A. –5 J/K
B. 0 J/K
C. 5 J/K
D. 10 J/K

69. Two ideal gases, A and B, are at the same temperature, volume and pressure. Gas A is reversibly expanded at constant temperature to a volume V. Gas B is allowed to expand into an evacuated chamber until it also has a total volume V, but without exchanging heat with its surroundings. Which of the following most accurately describes the two gases?

A. Gas A has a higher temperature and enthalpy than gas B.
B. Gas A has a higher temperature but a lower enthalpy than gas B.
C. Gas B has a higher temperature and enthalpy than gas A.
D. Gas A and B have equal temperatures and enthalpies.

STOP. IF YOU FINISH BEFORE TIME IS CALLED, CHECK YOUR WORK. YOU MAY GO BACK TO ANY QUESTION IN THIS TEST BOOKLET.

STOP.

30-minute In-class Exam for Lecture 4

Passage I (Questions 70-75)

The following tests were carried out on samples of an unknown solution in order to identify any presence of nitrate and nitrite ions.

Experiment 1

In acidic solution nitrites react with sulfamic acid (HNH_2SO_3) according to the following reaction:

$$NO_2^- + NH_2SO_3^- \rightarrow N_2(g) + SO_4^{2-} + H_2O$$

Barium sulfate is insoluble while barium sulfamate is soluble.

In a test tube, $BaCl_2$ is added to a few drops of a sample of the unknown solution. No precipitate is formed. A few crystals of sulfamic acid are then mixed into the sample.

No visible reaction occurs.

Experiment 2

Active metals such as aluminum and zinc in alkaline solution reduce nitrate to ammonia. Nitrite ion will also form ammonia under these conditions. Devarda's alloy (50% Cu, 45% Al, 5% Zn) gives the following reaction with nitrate ion:

$$Al + NO_3^- \rightarrow Al(OH)_4^- + NH_3(g)$$

Several drops of the unknown solution are mixed with an equal amount of 6 *M* NaOH and placed into a dry test tube. Care is taken not to wet the walls of the tube. Devarda's alloy is then added and a loose cotton plug pushed one third of the way down the tube. The tube is warmed briefly in a water bath and removed. A bent strip of red litmus with a moistened fold is then placed at the top of the tube as shone in Figure 1.

The moistened section of the litmus turns blue.

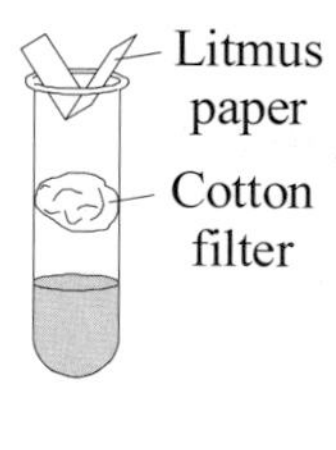

Figure 1

70. The unknown solution contains which of the following ions:

A. nitrite but not nitrate
B. both nitrite and nitrate
C. nitrate but not nitrite
D. neither nitrite nor nitrate

71. Which of the following is true of the oxidation state of aluminum in the reaction in Experiment 2?

A. Aluminum is reduced to +3.
B. Aluminum is oxidized to +3.
C. Aluminum is reduced to +4.
D. Aluminum is oxidized to +4.

72. If the unknown solution contained nitrite what would be the expected result of Experiment 1?

A. A precipitate would be formed before the addition of sulfamic acid.
B. Bubbles and precipitate would be observed after the addition of sulfamic acid.
C. Bubbles but no precipitate would be observed after the addition of sulfamic acid.
D. No visible reaction would be observed after the addition of sulfamic acid.

73. What is the approximate concentration of H^+ in the NaOH solution used in Experiment 2?

A. $1\text{x}10^{-15}$ *M*
B. $1.6\text{x}10^{-15}$ *M*
C. $6\text{x}10^{-14}$ *M*
D. 6 *M*

74. When the solution in Experiment 2 is warmed in the water bath, all of the following may be true EXCEPT:

A. The equilibrium constant *K* of the reaction changes.
B. The rate constant *k* of the reaction changes.
C. The rate of the reaction decreases.
D. The rate of the reaction increases.

75. Which of the following represents the reaction taking place in the litmus paper in Experiment 2?

A. $NO_3^- + H_2O \rightarrow OH^- + HNO_3$
B. $HNO_3 + H_2O \rightarrow H_3O^+ + NO_3^-$
C. $NH_4^+ + H_2O \rightarrow H_3O^+ + NH_3$
D. $NH_3 + H_2O \rightarrow NH_4^+ + OH^-$

GO ON TO THE NEXT PAGE.

Passage II (Questions 76-82)

When $Ca(IO_3)_2$ dissolves in a solution containing H^+ the following two reactions occur.

$$Ca(IO_3)_2 \rightleftharpoons Ca^{2+} + 2IO_3^-$$

Reaction 1

$$H^+ + IO_3^- \rightleftharpoons HIO_3$$

Reactions 2

HIO_3 is a weak acid. The K_{sp} for $Ca(IO_3)_2$ and the K_a for HIO_3 can be determined from the solubility *(S)* of $Ca(IO_3)_2$ for solutions of varying $[H^+]$. The solubility is related to the initial hydrogen ion concentration $[H^+]$ by the following equation:

$$2(\underline{S})^{3/2} = K_{sp}^{1/2} + \left[\frac{K_{sp}^{1/2}}{K_a}\right][H^+]$$

A student prepared four saturated solutions by mixing $Ca(IO_3)_2$ with a strong acid. Excess solid was filtered off. The student found the *S* for each solution with constant ionic strength, using iodometric titrations. The resulting data are shown in Table 1.

Solution	$[H^+]$, *M*	*(S)*, *M*
1	0	0.0325
2	0.260	0.0442
3	0.520	0.0527
4	0.781	0.0654

Table 1. Solubility data for $Ca(IO_3)_2$

76. The K_{sp} for $Ca(IO_3)_2$ and the K_a for HIO_3, respectively are:

A. $[Ca^{2+}][IO_3^-]^2$ and $\dfrac{[H^+][IO_3^-]}{[HIO_3]}$

B. $\dfrac{[Ca^{2+}][IO_3^-]^2}{[Ca(IO_3)_2]}$ and $\dfrac{[H^+][IO_3^-]}{[HIO_3]}$

C. $[Ca^{2+}][IO_3^-]^2$ and $[H^+][IO_3^-]$

D. $[Ca^{2+}][IO_3^-]^2$ and $\dfrac{[HIO_3]}{[H^+][IO_3^-]}$

77. As $[H^+]$ increases, the solubility of $Ca(IO_3)_2$:

A. increases and K_{sp} increases.
B. decreases and K_{sp} decreases.
C. increases and K_{sp} does not change.
D. does not change and K_{sp} increases.

78. The graph of $2(\underline{S})^{3/2}$ versus $[H^+]$ for the data shown in Table 1 would most closely resemble which of the following?

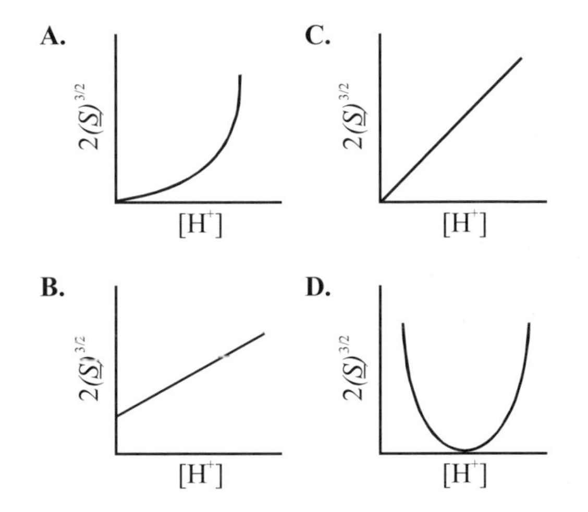

79. After filtering out excess solid, a student adds HCl to Solution 1 in Table 1. He then adds a small amount of $CaSO_4$, which dissolves completely. Which of the following also occurs in the new solution?

A. Some $Ca(IO_3)_2$ precipitates when the $CaSO_4$ is added.
B. Undissociated HIO_3 increases when the HCl is added.
C. Aqueous IO_3^- decreases when $CaSO_4$ is added.
D. Aqueous Ca^{2+} decreases when $CaSO_4$ is added.

80. If a strong acid were added after the filtration of Solution 4, the measured solubility of $Ca(IO_3)_2$ would:

A. increase.
B. decrease.
C. remain unchanged.
D. The measured solubility of $Ca(IO_3)_2$ cannot be determined.

GO ON TO THE NEXT PAGE.

81. If $Ca(OH)_2$ is added to the Solution 3 in Table 1:

A. the concentration of H^+ will increase.
B. the concentration of HIO_3 will increase.
C. the concentration of IO_3^- will increase
D. $Ca(IO_3)_2$ will precipitate.

82. How will the addition of HIO_3 affect Solution 2 from Table 1?

A. The lower pH will shift Reaction 2 to the right.
B. The increased hydrogen ion concentration will dissolve more $Ca(IO_3)_2$.
C. The common ion effect will shift Reaction 1 to the left.
D. The lower pH will balance out the common ion effect and the equilibrium will not change.

Passage III (Questions 83-89)

Many carbonate minerals are found in the earth's crust. As a result, the waters of several lakes, rivers, and even oceans are in contact with these minerals. $CaCO_3$ is the primary component of limestone and marble, while dolomite ($CaMg(CO_3)_2$) and magnesite ($MgCO_3$) are minerals found in other rock formations.

Limestone lines many of the river and lake beds resulting in contamination of the fresh water supply with Ca^{2+} and Mg^{2+}. The amount of these minerals present in water can be measured in parts per million (ppm). The "hardness" of water is determined by the ppm of Ca^{2+} and Mg^{2+} present. Hard water is the cause of many problems in the home. Scale buildup in pipes, on pots and pans, and in washing machines are just a few of the problems.

The hardness of water can be measured by titrating a sample of water with the ligand ethylenediamine tetraacetic acid (EDTA) and the indicator eriochrome black T. This ligand forms a coordination complex with metal cations in a one-to-one stochiometry. The structure of EDTA is shown in Figure 1. It has six binding sites to form a very stable complex ion with most metal ions.

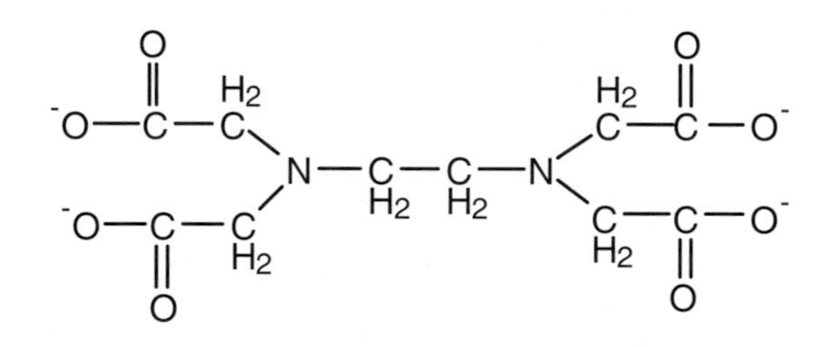

Figure 1 The structure of EDTA

The association constants for EDTA with several metal ions are listed in Table 1.

Metal	K_{assoc}
Ag^+	2×10^7
Mg^{2+}	5×10^8
Ca^{2+}	5×10^{10}
Al^{3+}	1×10^{16}
Fe^{2+}	2×10^{14}
Fe^{3+}	1×10^{25}
Cu^{2+}	6×10^{18}
Co^{2+}	2×10^{16}
Hg^{2+}	6×10^{21}

Table 1 Association constants for EDTA with metal ions

GO ON TO THE NEXT PAGE.

Many households soften hard water using ion exchange resins. These resins replace the Ca^{2+} and Mg^{2+} ions with smaller cations such as Na^+ and H^+.

83. A scientist determines how hard the tap water is in the laboratory, using an EDTA titration. If the pipes in the building are old and some rust dissolves into the tap water, how will the results of the test change?

A. The results will not change because the EDTA titration only works with Ca^{2+} and Mg^{2+}.
B. The titration will not be able to be carried out because the tap water will be colored.
C. The tap water will appear to have less Ca^{2+} and Mg^{2+} present.
D. The tap water will appear to have more Ca^{2+} and Mg^{2+} present.

84. When EDTA reacts with a metal ion to form a complex ion, EDTA is acting as a(n):

A. oxidizing agent.
B. reducing agent.
C. Lewis base.
D. Lewis acid.

85. Salt water contains a high concentration of Cl^- ions. These ions form complexes ($CaCl^+$) with Ca^{2+}. How will the solubility of limestone change in ocean water compared to fresh water?

A. It will increase.
B. It will decrease.
C. It will remain the same.
D. The change in solubility cannot be determined.

86. A 25 mL sample of hard water is titrated with a 0.001 *M* solution of EDTA, and the endpoint of the titration is reached at 50 mL of EDTA added. What is the concentration of Ca^{2+} and Mg^{2+} ions in solution?

A. 0.0005 *M*
B. 0.001 *M*
C. 0.002 *M*
D. 0.006 *M*

87. The EDTA titrations are carried out at a pH of 10. Why is it necessary to buffer the pH at 10?

A. A low pH will cause metal hydroxides to form.
B. $CaCO_3$ requires a high pH in order to dissolve.
C. The indicator requires pH 10 to change color.
D. The coordinating atoms must be deprotonated in order to bond with the metal ion.

88. Why does replacing the cations found in hard water with Na^+ or H^+ soften the water (i.e., reduce the unwanted residue produced by hard water)?

A. The smaller cations do not form insoluble mineral deposits.
B. Twice as many smaller ions are necessary to react with soaps and other ligands.
C. No minerals contain Na and H.
D. H is found in water so there is no addition of new atoms.

89. 9 ppm is equivalent to an aqueous concentration of approximately 5×10^{-4} mol/L. If a water sample were reduced from 18 ppm Mg^{2+} to 9 ppm Mg^{2+} by the addition of EDTA, according to Table 1 what would be the concentration of the remaining unbound EDTA?

A. 2×10^{-9} mol/L
B. 5×10^{-4} mol/L
C. 1×10^{-3} mol/L
D. 5×10^{-3} mol/L

GO ON TO THE NEXT PAGE.

Questions 90 through 92 are **NOT** based on a descriptive passage.

90. The vapor pressure of pure water at 25°C is approximately 23.8 torr. Which of the following is the vapor pressure of pure water at 95°C?

A. 10 torr
B. 23.8 torr
C. 633.9 torr
D. 800 torr

91. Benzene and toluene form a nearly ideal solution. If the vapor pressure for benzene and toluene at 25°C is 94 mm Hg and 29 mm Hg respectively, what is the approximate vapor pressure of a solution made from 25% benzene and 75% toluene at the same temperature?

A. 29 mm Hg
B. 45 mm Hg
C. 94 mm Hg
D. 123 mm Hg

92. When volatile solvents A and B are mixed in equal proportions heat is given off to the surroundings. If pure A has a higher boiling point than pure B, which of the following could NOT be true?

A. The boiling point of the mixture is less than pure A.
B. The boiling point of the mixture is less than pure B.
C. The vapor pressure of the mixture is less than pure A.
D. The vapor pressure of the mixture is less than pure B.

STOP. IF YOU FINISH BEFORE TIME IS CALLED, CHECK YOUR WORK. YOU MAY GO BACK TO ANY QUESTION IN THIS TEST BOOKLET.

STOP.

30-minute In-class Exam for Lecture 5

Passage I (Questions 93-100)

Two students performed an experiment using the apparatus shown in Figure 1.

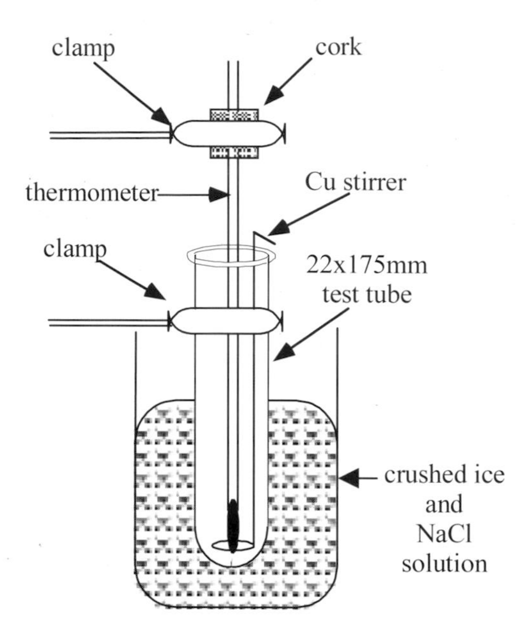

Figure 1. Freezing Point Apparatus

Both students placed 10.00 mL of cyclohexane into the test tube at room temperature. Next 0.500 gram of an unknown solid was dissolved in the cyclohexane. The test tube and contents were lowered into the ice bath, which was maintained at a temperature of –5.0°C by adjusting the relative amounts of NaCl, ice, and water. The students monitored the temperature of the cyclohexane mixtures by taking readings from the thermometer at 30 seconds intervals. The freezing point of the solution for a given trial is the temperature maintained for four consecutive readings. The experiment was repeated two more times by warming the cyclohexane to room temperature then freezing it again.

The results obtained by the students are recorded in Table 1.

Time, seconds	30	60	90	120	150	180	210	240
Student 1	22.0	6.0	4.0	-3.5	-3.5	-3.5	-3.5	-3.5
Student 2	22.0	12.0	6.0	0.6	0.6	0.6	0.6	0.4

Table 1. Solution temperature (°C) with time

(The K_f for cyclohexane is 20.2°C kg/mol, the freezing point is 6.6°C, and the density is 0.78 g/mL. The K_f for water is 1.86°C kg/mol.)

93. Why was NaCl added to the ice bath?

- **A.** To lower the freezing point of the water and cool the cyclohexane solution more quickly.
- **B.** To lower the freezing point of the cyclohexane solution below the freezing point of the water.
- **C.** To lower the freezing point of the water below the freezing point of the cyclohexane solution.
- **D.** To raise the freezing point of the water.

94. Which salt is the most efficient per gram at lowering the freezing point of water?

- **A.** $Ba(OH)_2$
- **B.** $MgSO_4$
- **C.** NaCl
- **D.** $CaCl_2$

95. The purpose of the copper stirrer is:

- **A.** to ensure that the solid stays in solution.
- **B.** to create heat to offset the chilling effect of the ice bath.
- **C.** to ensure that the solution temperature remains homogenous.
- **D.** to allow the student to see when crystals begin to form.

96. According to the results in Table 1, which student had the unknown with the greatest molecular weight? (Assume no dissociation of the unknown solids occurs.)

- **A.** Student 1
- **B.** Student 2
- **C.** The molecular weights were the same.
- **D.** It cannot be determined based on the given information.

GO ON TO THE NEXT PAGE.

97. If the unknowns were soluble in water, and water were used as the solvent instead of cyclohexane, the freezing point of the aqueous solutions in both experiments would be:

A. lower than both cyclohexane solutions.
B. higher than both cyclohexane solutions.
C. lower than the freezing point of the cyclohexane solution used by Student 1, but higher than the cyclohexane solution used by Student 2.
D. higher than the freezing point of the cyclohexane solution used by Student 1, but lower than the cyclohexane solution used by Student 2.

98. If no dissociation of the unknown solid occurred, the molecular weight of the unknown used by Student 1 is:

A. $\dfrac{(0.5)\times(20.2)}{(0.78)\times(10)\times(10.1)}$

B. $\dfrac{(0.5)\times(20.2)\times(1000)}{(0.78)\times(10)\times(3.5)}$

C. $\dfrac{(0.5)\times(20.2)\times(1000)}{(0.78)\times(10)\times(6.0)}$

D. $\dfrac{(0.5)\times(20.2)\times(1000)}{(0.78)\times(10)\times(10.1)}$

99. A professor must choose the unknown from the following solutes. Which would be the best solute to give to his students for the experiment in the passage?

A. NaCl
B. $Mg(OH)_2$
C. CH_3OH
D. $C_{10}H_8$

100. Why does the temperature in Student 2's experiment begin to drop after 210 seconds?

A. Student 2 used too much ice in the ice bath.
B. Student 2's cyclohexane solution was completely frozen at 210 seconds.
C. Student 2 stopped stirring the solution at 180 seconds.
D. Student 2 stopped dissolving the unknown solid.

Passage II (Questions 101-106)

A series of experiments are performed using the calorimeter shown in Figure 1.

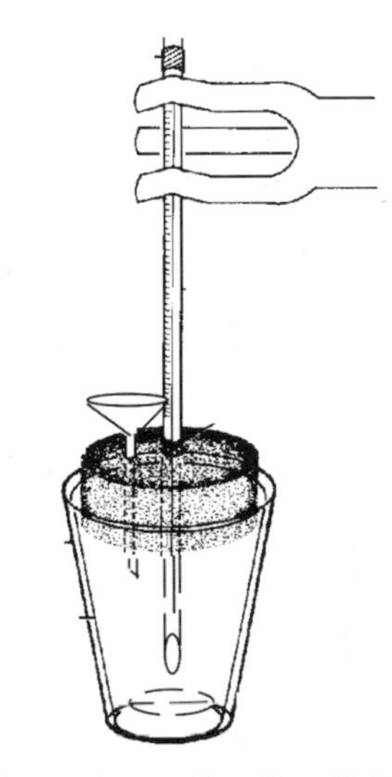

Figure 1. Coffee Cup Calorimeter

A volume of 0.5 *M* NaOH is placed near the calorimeter, which contains an equal volume of 0.5 *M* HCl. The temperatures of both solutions are monitored until they equilibrate to room temperature.

The NaOH solution is added to the HCl solution through the funnel. The temperature is recorded every 30 seconds for 5 minutes. The experiment is repeated three times with three different volumes of HCl and NaOH. The results of one of these experiments are shown in the graph in Figure 2. The data for all experiments are recorded in Table 1.

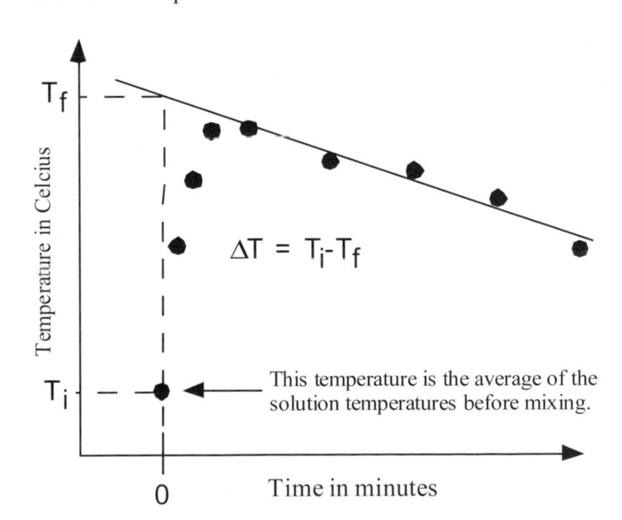

Figure 2 Temperature change of solution over time

GO ON TO THE NEXT PAGE.

Trial	Volume of HCl and NaOH (mL)	Initial Temperature (°C)	Final Temperature (°C)
1	30	22.0	25.3
2	40	20.0	23.3
3	50	21.0	24.3

Table 1

101. What reaction is taking place in the calorimeter to cause the temperature change?

A. $H^+ + OH^- \rightarrow H_2O$
B. $Na^+ + Cl^- \rightarrow NaCl$
C. $NaCl \rightarrow Na^+ + Cl^-$
D. $Na^+ + 1e^- \rightarrow Na$

102. The reaction in the calorimeter is an:

A. endothermic reaction.
B. exothermic reaction.
C. oxidation reaction.
D. isothermic reaction.

103. Assuming that the heat capacity of the solution is the same as the heat capacity of water, what is the enthalpy change for the reaction in Trial 2 as recorded in Table 1? (The heat capacity for water is 1.0 cal $°C^{-1}$ mL^{-1})

A. –132 cal
B. 132 cal
C. –330 cal
D. 330 cal

104. If 0.5 *M* NH_4OH (a weaker base) were used instead of NaOH, how would this affect the results of the experiment?

A. The temperature change would be greater because more energy is required to dissociate NH_4OH.
B. The temperature change would be less because more energy is required to dissociate NH_4OH.
C. The temperature change would be greater because less energy is required to dissociate NH_4OH.
D. It would not change the results because both bases are ionic compounds and the energy required to separate equal charges is always the same.

105. Which trial gave the greatest heat of solution per mole of reactants?

A. 1
B. 2
C. 3
D. They should all be the same.

106. If the solutions in the experiment began at room temperature, which of the following explains the heat transfer between the calorimeter and its surroundings for the experiment shown in Figure 2?

A. Initially heat is transferred from the surroundings to the calorimeter, and then heat is transferred from the calorimeter to the surroundings.
B. Initially heat is transferred from the calorimeter to the surroundings, and then heat is transferred from the surroundings to the calorimeter.
C. Heat is transferred from the surroundings to the calorimeter throughout the experiment.
D. Heat is transferred from the calorimeter to the surroundings throughout the experiment.

GO ON TO THE NEXT PAGE.

Passage III (Questions 107-113)

Phase diagrams show the changes in phase of a material as a function of temperature and pressure. Student A prepared a phase diagram for CO_2. After observing the phase diagram, he concluded that raising the pressure isothermally promotes a substance to change from a gas to a liquid to a solid as demonstrated by the dashed line in Figure 1.

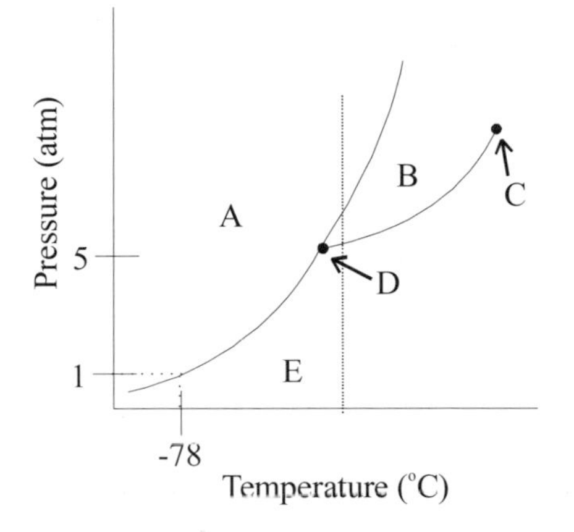

Figure 1 Phase diagram of CO_2

Student B chose to make a phase diagram of H_2O. She observed that raising the pressure isothermally promotes a substance to convert from vapor to solid then to liquid as indicated by the dashed line in Figure 2.

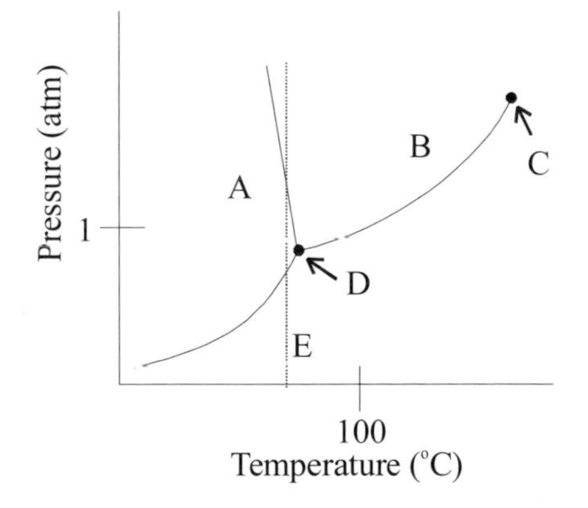

Figure 2 Phase diagram of H_2O

107. Which of the following explains the discrepancy between the observations of the two students?

A. Water expands when going from liquid to solid, where as CO_2 contracts.
B. CO_2 expands when going from liquid to solid, where as water contracts.
C. The two chemists observed the phase changes at different temperatures.
D. CO_2 is a gas at room temperature, while H_2O is a liquid.

108. According to Figure 1, at –78°C and 1 atm CO_2 will:

A. exist as a liquid.
B. exist in equilibrium as a gas and liquid.
C. exist in equilibrium as a gas and solid.
D. exist in equilibrium as a liquid and solid.

109. Which of the following is true concerning CO_2 at point D in Figure 1?

A. It is a vapor.
B. It is a liquid.
C. It is a solid.
D. All three phases can exist in equilibrium.

110. At temperatures and pressures greater than point C in Figure 1:

A. CO_2 is a vapor.
B. CO_2 is a liquid.
C. CO_2 is in both liquid and vapor phase.
D. the vapor and liquid phases of CO_2 cannot be distinguished.

111. According to Figure 2, as the pressure increases the melting point of H_2O?

A. increases
B. decreases
C. does not change
D. increases than decreases

112. The normal boiling point for O_2 is 90.2 K. Which of the following could be the triple point for O_2?

A. 1.14 mmHg and 54.4 K
B. 1.14 mmHg and 154.6 K
C. 800 mmHg and 54.4 K
D. 37,800 mmHg and 154.6 K

113. Describe the phase change for H_2O as the pressure is raised at 100°C.

A. sublimation
B. vaporization
C. condensation
D. melting

GO ON TO THE NEXT PAGE.

Questions 114 through 115 are **NOT** based on a descriptive passage.

114. During a solid to liquid phase change, energy is:

A. absorbed by bond breakage.
B. released by bond breakage.
C. absorbed by increased kinetic energy of the liquid molecules.
D. released by increased kinetic energy of the liquid molecules.

115. On his honeymoon the chemist, Joule, took with him a long thermometer with which to measure the temperature difference between the waters at the top and the bottom of Niagra Falls. If the height of the falls is 60 meters and the specific heat of water is approximately 4200 J kg^{-1} K^{-1}, what is the expected temperature difference?

A. 1/7 K
B. 7 K
C. 70 K
D. 700 K

STOP. IF YOU FINISH BEFORE TIME IS CALLED, CHECK YOUR WORK. YOU MAY GO BACK TO ANY QUESTION IN THIS TEST BOOKLET.

STOP.

30-minute In-class Exam for Lecture 6

Passage I (Questions 116-121)

The solubility of $Ca(OH)_2$ (Reaction 1) can be determined by titrating the saturated solution containing no precipitate against a standardized HCl solution and determining $[OH^-]$.

$$Ca(OH)_2 \rightleftharpoons Ca^{2+} + 2OH^-$$

Reaction 1

Once $[OH^-]$ is determined, the solubility *(S)* of $Ca(OH)_2$ is calculated using the following equation:

$$S = {}^1/_2[OH^-]_{Ca(OH)_2}$$

Equation 1

where $[OH^-]_{Ca(OH)_2}$ is the concentration of hydroxide ion due only to $Ca(OH)_2$. The solubilities of $Ca(OH)_2$ in a variety of solutions of varying $[OH^-]$ concentrations were determined by the above method, but the calculation of *S* had to be altered slightly due to the presence of additional hydroxide ions.

$$S = {}^1/_2\{[OH^-]_{total} - [OH^-]_{solvent}\}$$

Equation 2

The results of the experiment are summarized in Table 1.

Trial	Solution	Solubility
1	H_2O	0.0199 *M*
2	0.01793 *M* NaOH	0.0100 *M*
3	0.03614 *M* NaOH	0.0047 *M*
4	0.07119 *M* NaOH	0.0015 *M*

Table 1. Solubility data for $Ca(OH)_2$.

116. How does the solubility of $Ca(OH)_2$ change as the $[OH^-]$ in the solvent increases?

- **A.** It decreases because the increase in OH^- shifts Reaction 1 toward the left.
- **B.** It decreases because the increase in OH^- interferes with the acid titration.
- **C.** It increases because the increase in OH^- shifts Reaction 1 toward the left.
- **D.** It increases because the increase in OH^- interferes with the acid titration.

117. How do the titrations in Trials 1 and 3 compare?

- **A.** The pH of the equivalence points are the same, but more HCl is required to reach the equivalence point in Trial 3.
- **B.** The pH of the equivalence point in Trial 1 is higher, and less HCl is required to reach it.
- **C.** The pH of the equivalence point in Trial 3 is higher, and less HCl is required to reach it.
- **D.** The pH of the equivalence point in Trial 3 is higher, and more HCl is required to reach it.

118. The K_{sp} for Reaction 1 in the presence of NaOH is:

- **A.** $[Ca^{2+}][OH^-]^2_{Ca(OH)_2}$
- **B.** $[Ca^{2+}][OH^-]^2_{Total}$
- **C.** $[Ca^{2+}][2OH^-]^2_{Ca(OH)_2}$
- **D.** $[Ca^{2+}][2OH^-]^2_{Total}$

119. What is the pH of the solution in Trial 3 before the titration?

- **A.** 1.1
- **B.** 7.0
- **C.** 9.3
- **D.** 12.7

120. Which indicator would be best for the titration in this experiment?

- **A.** Phenolphthalein: (acid color is colorless, base is red, and the transition pH is 8.0 - 9.6).
- **B.** Thymolphthalein: (acid color is colorless, base is blue, and the transition pH is 8.3 - 10.5).
- **C.** Bromocresol purple: (acid color is yellow, base is purple, and the transition pH is 5.2 - 6.8).
- **D.** Neutral red: (acid color is red, base color is yellow, and the transition pH is 6.8 - 8.0).

GO ON TO THE NEXT PAGE.

121. If a pH meter were placed into the titration beaker, what would be the resulting curve for Trial 1?

A.

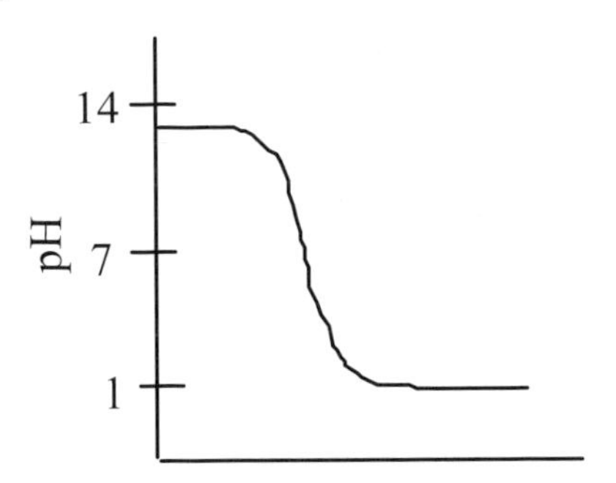

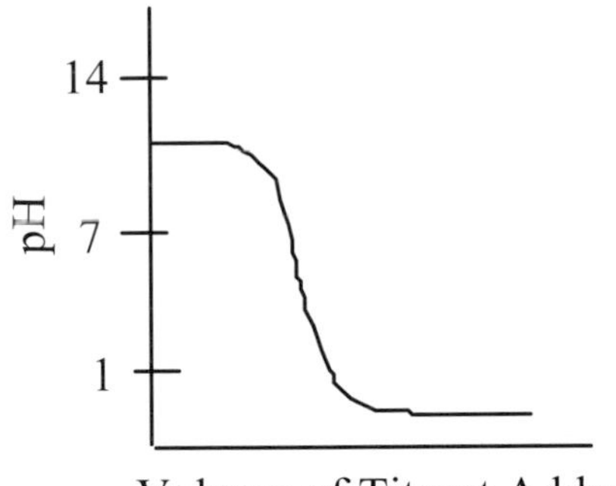

C.

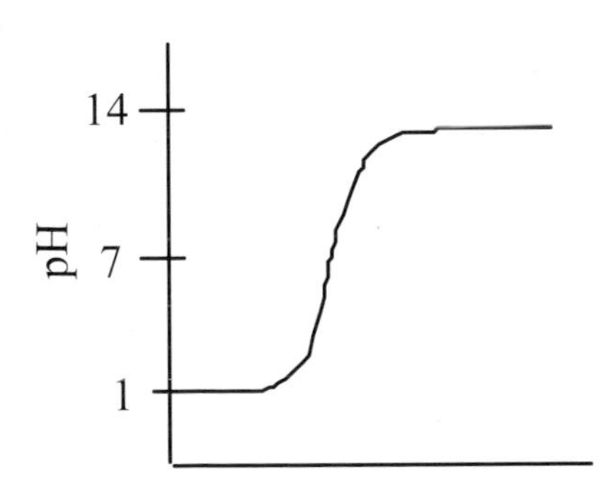

Volume of Titrant Added

D.

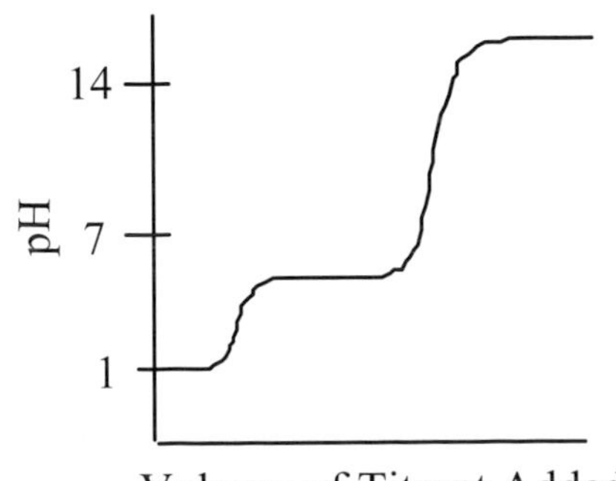

Passage II (Questions 122-127)

The reaction for the autoionization of water is shown below:

$$2H_2O \rightarrow H_3O^+ + OH^-$$

The equilibrium constant (K_w) is temperature dependent. Table 1 lists the value of K_w at several temperatures.

Temperature (°C)	K_w
0	0.114 x 10^{-14}
10	0.292 x 10^{-14}
20	0.681x 10^{-14}
25	1.01 x 10^{-14}
30	1.47 x 10^{-14}
40	2.92 x 10^{-14}
50	5.47 x 10^{-14}
60	9.61 x 10^{-14}

Table 1 Equilibrium constants for water at different temperatures

Water has a leveling effect on acids. Any acid stronger than H_3O^+ appears to have the same behavior in aqueous solution. For example, 1*M* HCl and 1*M* $HClO_4$ have the same concentration of H_3O^+ even though in anhydrous acetic acid, $HClO_4$ is a stronger acid.

122. What is the pH of H_2O at 40°C?

A. 7.5
B. 7.0
C. 6.7
D. 6.0

123. At 10 °C, the concentration of OH^- in 1 *M* HCl is approximately:

A. 1 x 10^{-7} *M*
B. 1 x 10^{-14} *M*
C. 3 x 10^{-15} *M*
D. 1 x 10^{-15} *M*

124. As temperature increases, the pH of pure water:

A. increases.
B. decreases.
C. becomes less than the pOH
D. becomes greater than the pOH.

GO ON TO THE NEXT PAGE.

125. What is the conjugate base of H_2SO_4?

A. H_2O
B. OH^-
C. HSO_4^-
D. SO_4^{2-}

126. Why can the relative strength of HCl and $HClO_4$ be determined in acetic acid but not in water?

A. because acetic acid is a weaker acid than H_3O^+
B. because acetic acid is a stronger acid than H_3O^+
C. because acetic acid is a weaker Bronsted-Lowry base than H_2O
D. because acetic acid is a stronger Bronsted-Lowry base than H_2O

127. The equation for K_w at 50 °C is:

A. $[OH^-][H_3O^+]$

B. $\dfrac{[OH^-][H_3O^+]}{[H_2O]^2}$

C. $\dfrac{[OH^-][H_3O^+]}{[H_2O]}$

D. $[H_3O^+]$

Passage III (Questions 128-1)

Acid rain results when $SO_3(g)$, produced by the industrial burning of fuel, dissolves in the moist atmosphere.

$$SO_3(g) + H_2O\ (l) \rightarrow H_2SO_4(aq)$$

The rain formed from the condensation of this acidic water is an environmental hazard destroying trees and killing the fish in some lakes. (The pH of the water varies depending upon the level of pollution in the area. The pK_a values are about –2 for H_2SO_4 and 1.92 for HSO_4^-.)

Another pollutant which dissolves in water vapor and reacts to form acid rain is $SO_2(g)$. This gas forms $H_2SO_3(aq)$ which can be oxidized to H_2SO_4. (The pK_a values are 1.81 for H_2SO_3 and 6.91 for HSO_3^-.)

The table below gives the color changes of many acid base indicators used to test the pH of water.

Indicator	Color Change	pH of color change
Malachite green	yellow to green	0.2-1.8
Thymol blue	red to yellow	1.2-2.8
Methyl orange	red to yellow	3.2-4.4
Methyl red	red to yellow	4.8-6.0
Phenolphthalein	clear to red	8.2-10.0
Alizarin yellow	yellow to red	10.1-12.0

Table 1

128. A sample of rainwater tested with methyl orange results in a yellow color, and the addition of methyl red to a fresh sample of the same water results in a red color. What is the pH of the sample?

A. between 1.2 and 1.8
B. between 3.2 and 4.4
C. between 4.4 and 4.8
D. between 4.8 and 6.0

GO ON TO THE NEXT PAGE.

129. If there is no oxidant present in the air and the same number of moles of SO_2 and SO_3 are dissolved, which gas would produce acid rain with a lower pH?

A. SO_2 because H_2SO_3 has a higher pK_a than H_2SO_4.
B. SO_2 because HSO_3^- has a higher pK_a than HSO_4^-.
C. SO_3 because H_2SO_4 has a lower pK_a than H_2SO_3.
D. SO_3 because HSO_4^- has a lower pK_a than HSO_3^-.

130. The pH of which of the following lakes would be most affected by acid rain?

A. a large shallow lake
B. a small deep lake
C. a large shallow lake lined with limestone ($CaCO_3$)
D. a small deep lake lined with limestone ($CaCO_3$)

131. What is the conjugate base of HSO_3^-?

A. OH^-
B. H^+
C. SO_3^{2-}
D. H_2SO_3

132. What is the oxidation state of sulfur in H_2SO_4 and H_2SO_3 respectively?

A. +6, +4
B. +4, +6
C. –6, –4
D. –4, –6

133. What is the pH of a 5.0 x 10^{-8} M aqueous solution of H_2SO_4 at room temperature?

A. 8.3
B. 7.3
C. 6.8
D. 6.0

134. A sample of rainwater polluted with SO_3 is titrated with NaOH. Which of the following most resembles the shape of titration curve.

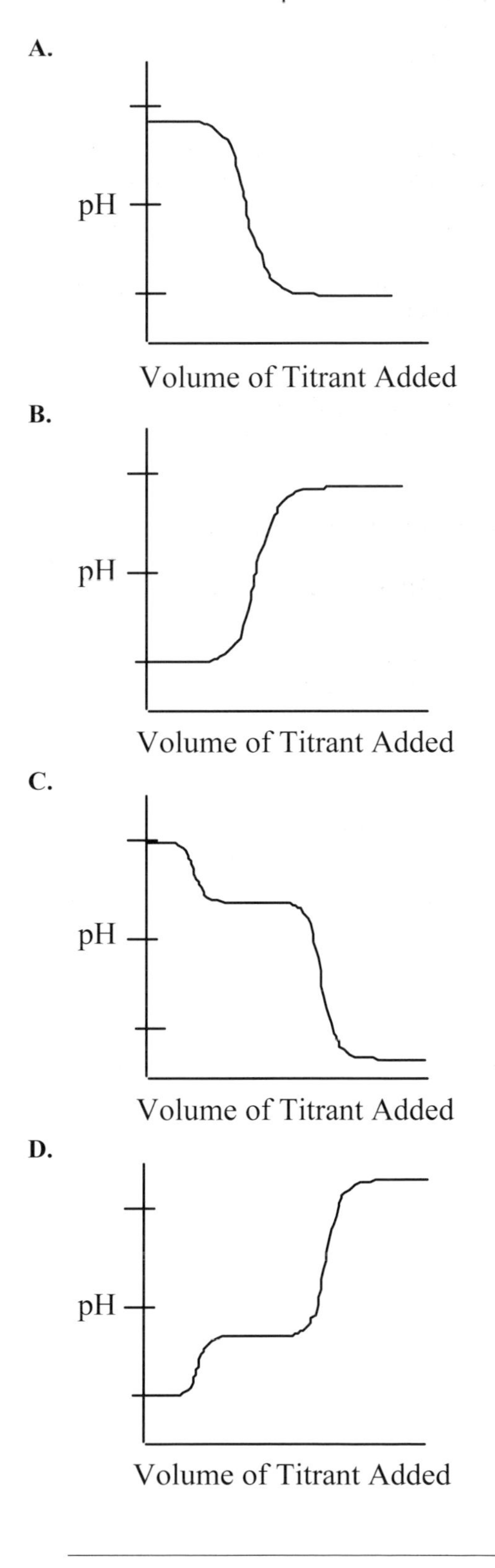

GO ON TO THE NEXT PAGE.

Questions 135 through 138 are **NOT** based on a descriptive passage.

135. Which of the following is the strongest base?

A. ClO^-
B. ClO_2^-
C. ClO_3^-
D. ClO_4^-

136. A weak acid is titrated with a strong base. When the concentration of the conjugate base is equal to the concentration of the acid, the titration is at the:

A. stoichiometric point.
B. equivalence point.
C. half equivalence point.
D. end point.

137. A buffer solution is created using acetic acid and its conjugate base. If the ratio of acetic acid to its conjugate base is 10 to 1, what is the approximate pH of the solution? (The K_a of acetic acid is $1.8\text{x}10^{-5}$)

A. 3.7
B. 4.7
C. 5.7
D. 7.0

138. NH_3 has a K_b of $1.8\text{x}10^{-5}$. Which of the following has a K_a of $5.6\text{x}10^{-10}$?

A. NH_3
B. NH_4^+
C. NH_2^-
D. H^+

STOP. IF YOU FINISH BEFORE TIME IS CALLED, CHECK YOUR WORK. YOU MAY GO BACK TO ANY QUESTION IN THIS TEST BOOKLET.

STOP.

30-minute In-class Exam for Lecture 7

Passage I (Questions 139-144)

When Fe^{2+} is titrated with dichromate ($Cr_2O_7^{2-}$) according to Reaction 1, the titration curve similar to the one shown in Figure 1 results. The curve was generated by measuring the potential difference between the reaction solution and a standard solution after each addition of a known volume and concentration of dichromate. The potential is measured using a voltmeter attached to an *orp* electrode.

$$6Fe^{2+} + Cr_2O_7^{2-} + 14H^+ \rightleftharpoons 2Cr^{3+} + 7H_2O + 6Fe^{3+}$$

Reaction 1

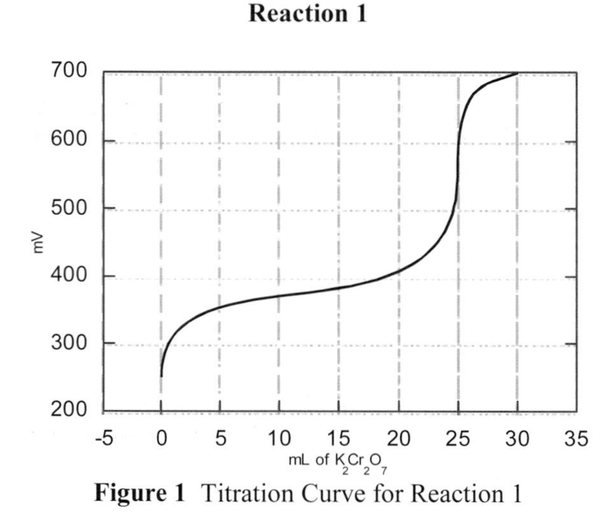

Figure 1 Titration Curve for Reaction 1

The endpoint of the titration can be indicated by the redox indicator diphenylamine sulfonic acid (DAS). However, the formal potential of the Fe solution must be lowered in order to match the endpoint with the equivalence point of the titration. This is accomplished by the addition of H_2SO_4 and H_3PO_4 immediately before titration.

139. Which of the following is true concerning Reaction 1?

A. Cr is oxidized and $Cr_2O_7^{2-}$ is the oxidizing agent.
B. Cr is reduced and $Cr_2O_7^{2-}$ is the reducing agent.
C. Cr is reduced and $Cr_2O_7^{2-}$ is the oxidizing agent.
D. Cr is oxidized and $Cr_2O_7^{2-}$ is the reducing agent.

140. What is the volume of dichromate added to the titration at the equivalence point?

A. 13 mL
B. 15 mL
C. 25 mL
D. 30 mL

141. Which of the following expressions gives the concentration of Fe^{2+} in the unknown solution in terms of the volume of dichromate added $V_{Cr_2O_7^{2-}}$, the molarity of dichromate $M_{Cr_2O_7^{2-}}$, and the original volume of Fe^{2+} solution V_{Fe}?

A. $\dfrac{(V_{Cr_2O_7^{2-}})(M_{Cr_2O_7^{2-}})}{V_{Fe}}$

B. $\dfrac{(6V_{Cr_2O_7^{2-}})(M_{Cr_2O_7^{2-}})}{V_{Fe}}$

C. $\dfrac{(V_{Cr_2O_7^{2-}})(M_{Cr_2O_7^{2-}})}{6V_{Fe}}$

D. $\dfrac{(2V_{Cr_2O_7^{2-}})(M_{Cr_2O_7^{2-}})}{V_{Fe}}$

142. Why were H_2SO_4 and H_3PO_4 added to the titration?

A. H^+ is involved in the reaction.
B. All redox titrations require the presence of acids.
C. H^+ is required to dissolve Fe.
D. The formal potential of the Fe solution needed to be lowered.

143. At the equivalence point of the titration:

A. the number of electrons received by Cr is equal to the number of electrons lost from Fe.
B. the number of electrons received by Fe is equal to the number of electrons lost from Cr.
C. the number of moles of Fe is equal to the number of moles of Cr.
D. the indicator always changes color.

144. What is the oxidation state of Cr in $Cr_2O_7^{2-}$?

A. +12
B. +7
C. +6
D. +3

GO ON TO THE NEXT PAGE.

Passage II (Questions 145-151)

Rechargeable batteries have become an essential part of our environmentally conscientious society. The nickel-cadmium cell battery is a rechargeable battery used in small electronic devices. The half reactions that take place in the nickel-cadmium battery during discharge are:

$$Cd(OH)_2(s) + 2e^- \rightarrow Cd(s) + 2OH^-$$

$$\mathscr{E}^o = -0.4\text{ V}$$

Half Reaction 1

$$2NiO_2(s) + H_2O + 2e^- \rightarrow 2Ni(OH)_2(s) + 2OH^-$$

$$\mathscr{E}^o = 0.5\text{ V}$$

Half Reaction 2

Other types of rechargeable batteries currently being developed are those using sodium or lithium metal as the anode and sulfur as the cathode. These batteries must operate at high temperatures because the metals must be in the liquid state, but they provide a high energy density, which means the batteries will be very light weight.

145. The reaction taking place at the anode when the nickel-cadmium batteries are discharging is:

A. $Cd(s) + 2OH^- \rightarrow Cd(OH)_2(s) + 2e^-$
B. $2NiO_2(s) + H_2O + 2e^- \rightarrow 2Ni(OH)_2(s) + 2OH^-$
C. $Cd(OH)_2(s) + 2e^- \rightarrow Cd(s) + 2OH^-$
D. $2Ni(OH)_2(s) + 2OH^- \rightarrow 2NiO_2(s) + H_2O + 2e^-$

146. When the nickel-cadmium battery is recharging, what is the reaction at the anode?

A. $Cd(s) + 2OH^- \rightarrow Cd(OH)_2(s) + 2e^-$
B. $2NiO_2(s) + H_2O + 2e^- \rightarrow 2Ni(OH)_2(s) + 2OH^-$
C. $Cd(OH)_2(s) + 2e^- \rightarrow Cd(s) + 2OH^-$
D. $2Ni(OH)_2(s) + 2OH^- \rightarrow 2NiO_2(s) + H_2O + 2e^-$

147. What is the oxidizing agent in the nickel cadmium battery during discharge?

A. Cd
B. $Cd(OH)_2$
C. NiO_2
D. $2Ni(OH)_2$

148. Which of the following is true concerning the nickel-cadmium battery when it is recharging?

A. The cell is a nonspontaneous electrolytic cell.
B. The cell is a nonspontaneous galvanic cell.
C. The cell is a spontaneous electrolytic cell.
D. The cell is a spontaneous galvanic cell.

149. In order to recharge the nickel-cadmium battery back to standard conditions, what is the minimum voltage that must be applied across its electrodes?

A. 0.1 V
B. 0.2 V
C. 0.9 V
D. 1.8 V

150. In a sodium-sulfur battery, what is the half reaction for sodium in the spontaneous direction?

A. $Na^+ + e^- \rightarrow Na$
B. $Na \rightarrow Na^+ + e^-$
C. $Na \rightarrow Na^{2+} + 2e^-$
D. $Na^+ + OH^- \rightarrow NaOH$

151. The nickel-cadmium battery is used to power a light bulb. The current in the light bulb flows:

A. in the same direction as the flow of electrons, from the side with Half Reaction 1 to the side with Half Reaction 2.
B. in the same direction as the flow of electrons, from the side with Half Reaction 2 to the side with Half Reaction 1.
C. in the opposite direction to the flow of electrons, from the side with Half Reaction 1 to the side with Half Reaction 2.
D. in the opposite direction to the flow of electrons, from the side with Half Reaction 2 to the side with Half Reaction 1.

GO ON TO THE NEXT PAGE.

Passage III (Questions 152-158)

A pH meter is a concentration cell which measures the potential difference between a reference solution and a test solution and reports the difference in terms of pH. In a simplified version of a pH meter the half reactions are:

$$H_2 + 2H_2O \rightarrow 2H_3O^+ + 2e^-$$
$$2H_3O^+ + 2e^- \rightarrow H_2 + 2H_2O$$

The potential difference between the two solutions is derived from the Nernst equation as follows:

$$\mathscr{E} = -\frac{0.0592}{2}\log\left(\frac{[H_3O^+]_{\text{test solution}}}{[H_3O^+]_{\text{reference solution}}}\right)^2$$

where E is given in volts. This equation can be rewritten in terms of the pH of the solutions as follows:

$$\mathscr{E} = 0.059(\text{pH}_{\text{test}} - \text{pH}_{\text{reference}})$$

Because it is inconvenient to bubble H_2 gas through a solution, a more sophisticated pH meter is used in standard laboratory practice. Dilute hydrochloric acid is used as the reference solution. The test solution is in contact with a thin glass membrane in which a silver wire coated with silver chloride is imbedded. This glass membrane is dipped into the test solution and the potential difference between the solutions is measured and interpreted by a computer, which displays the pH of the test solution. The same equation holds for both pH meters.

152. What is the reaction quotient (Q) in the Nernst equation for the simple pH meter?

A. $\dfrac{[H_3O^+]^2_{\text{test solution}}}{[H_2]}$

B. $\dfrac{[H_2]}{[H_3O^+]^2_{\text{test solution}}}$

C. $\dfrac{[H_3O^+]^2_{\text{test solution}}[H_2]_{\text{reference solution}}}{[H_3O^+]^2_{\text{reference solution}}[H_2]_{\text{test solution}}}$

D. $\dfrac{[H_3O^+]^2_{\text{test solution}}}{[H_3O^+]^2_{\text{reference solution}}}$

153. What would be the approximate potential difference measured by a pH meter if the test solution had a pH of 2 and the reference solution had a pH of 4?

A. –118 mV
B. –59 mV
C. 59 mV
D. 118 mV

154. The potential difference measured by a pH meter is directly proportional to:

A. the difference in the hydrogen ion concentrations of the test and reference solution.
B. the difference in the pH of the test and reference solution.
C. the pH of the test solution.
D. the hydrogen ion concentration of the test solution.

155. If the reference solution of a pH meter were 1 *M* HCl, and the potential difference measured by the meter were 59 mV, what would be the pH of the test solution?

A. 0
B. 1
C. 2
D. 8

156. How would the potential difference registered by a pH meter change for a given test solution if the hydrogen ion concentration of the reference solution were increased by a factor of 10?

A. The potential difference would increase by 59 mV.
B. The potential difference would decrease by 59 mV.
C. The potential difference would increase by a factor of 10.
D. The potential difference would decrease by a factor of 10.

GO ON TO THE NEXT PAGE.

157. A galvanic cell is prepared by connecting two half cells with a salt bridge and a wire. One cell has a Cu electrode and 1 *M* $CuSO_4$, and the other has a Cu electrode and 2 *M* $CuSO_4$. Which direction will the current flow through the wire?

A. toward the 1M $CuSO_4$ solution
B. toward the 2M $CuSO_4$ solution
C. current will not flow because the half reactions are the same for both sides.
D. current will not flow because both half cells have Cu electrodes.

158. Which of the following is true for an acid-base concentration cell such as the one used by the pH meter?

A. Current always flows toward the more acidic solution.
B. Current always flows toward the more basic solution.
C. Current always flows toward the more neutral solution.
D. Current always flows away from the more neutral solution.

Questions 159 through 161 are **NOT** based on a descriptive passage.

159. Consider the reduction potential:

$$Zn^{2+} + 2e^- \rightarrow Zn(s) \quad \mathcal{E}^o = -0.76 \text{ V}.$$

When solid Zinc is added to aqueous HCl, under standard conditions, does a reaction take place?

A. No, because the oxidation potential for Cl is positive.
B. No, because the reduction potential for Cl is negative.
C. Yes, because the reduction potential for H^+ is positive.
D. Yes, because the reduction potential for H^+ is zero.

160. Chemicals are mixed in a redox reaction and allowed to come to equilibrium. Which of the following must be true concerning the solution at equilibrium?

A. $K = 1$
B. $\Delta G^o = 0$
C. $\mathcal{E} = 0$
D. $\Delta G^o = \Delta G$

161. At 298 K all reactants and products in a certain oxidation-reduction reaction are in aqueous phase at initial concentrations of 1 *M*. If the total potential for the reaction is $\mathcal{E} = 20$ mV, which of the following must be true?

A. $K = 1$
B. $\mathcal{E}^o_{298} = 20$ mV
C. ΔG is positive.
D. $K < 1$

STOP. IF YOU FINISH BEFORE TIME IS CALLED, CHECK YOUR WORK. YOU MAY GO BACK TO ANY QUESTION IN THIS TEST BOOKLET.

STOP.

Answers

Lecture 1

1. D
2. B
3. A
4. B
5. C
6. D
7. B
8. C
9. D
10. C
11. D
12. C
13. C
14. A
15. A
16. B
17. B
18. D
19. B
20. A
21. D
22. B
23. D

Lecture 2

24. B
25. A
26. C
27. C
28. B
29. C
30. B
31. D
32. A
33. A
34. D
35. D
36. C
37. B
38. A
39. A
40. D
41. B
42. D
43. C
44. B
45. A
46. C

Lecture 3

47. A
48. C
49. B
50. C
51. A
52. D
53. C
54. C
55. D
56. C
57. C
58. A
59. B
60. C
61. D
62. A
63. A
64. B
65. A
66. B
67. C
68. C
69. D

Lecture 4

70. C
71. B
72. B
73. B
74. C
75. D
76. A
77. C
78. B
79. B
80. C
81. D
82. C
83. D
84. C
85. A
86. C
87. D
88. A
89. A
90. C
91. B
92. B

Lecture 5

93. C
94. C
95. C
96. B
97. D
98. D
99. D
100. B
101. A
102. B
103. A
104. B
105. D
106. D
107. A
108. C
109. D
110. D
111. B
112. A
113. C
114. A
115. A

Lecture 6

116. A
117. A
118. B
119. D
120. D
121. A
122. C
123. C
124. B
125. C
126. C
127. A
128. C
129. C
130. A
131. C
132. A
133. C
134. D
135. A
136. C
137. A
138. B

Lecture 7

139. C
140. C
141. B
142. D
143. A
144. C
145. A
146. D
147. C
148. A
149. C
150. B
151. D
152. D
153. A
154. B
155. B
156. A
157. A
158. B
159. D
160. C
161. B

Physical Sciences	
Raw Score	Estimated Scaled Score
23	15
22	14
21	13
19-20	12
18	11
16-17	10
15	9
13-14	8
12	7
10-11	6
9	5
7-8	4

Explanations to Chemistry In-class Exam 1

Passage I

1. **D is correct.** If you get a boiling point question on the MCAT, look for hydrogen bonding. It increases the strength of intermolecular attractions. Stronger intermolecular attractions leads to higher boiling point.
2. **B is correct.** You should recognize this compound as ionic because alkaline earth metals like to form ionic compounds with halogens.
3. **A is correct.** In order to explain an increase in boiling point, we have to look for a reason that intermolecular bond strength would increase. The intermolecular bonds in noble gases are totally due to Van der Waals forces. If the atoms are more polarizable, instantaneous dipoles can have greater strength. Larger atoms are more polarizable the electrons can get farther from the nucleus and create a larger dipole moment.
4. **B is correct.** This is a periodic trend. Radius increases going down and to the left on the periodic table.
5. **C is correct.** Crystallization depends upon molecular symmetry as well as intermolecular bonding. Boiling point is strongly dependent upon intermolecular bond strength.
6. **D is correct.** Methane is nonpolar, so its only intermolecular bonding is through van der Waals forces.
7. **B is correct.** All intermolecular bonding is via electrostatic forces. The dipoles in van der Waals forces are temporal whereas dipole-dipole interactions may be due to permanent dipoles.

Passage II

8. **C is correct.** 'As' is just to the left of 'Se' on the periodic table. Therefore, its radius should be slightly larger than Se.
9. **D is correct.** Elements in the same family tend to be chemically similar. Hydrogen is an exception.
10. **C is correct.** Atomic radius is a periodic trend increasing down and to the left.
11. **D is correct.** Only D is a true statement. A is knowledge that would not be required by the MCAT.
12. **C is correct.** Electron affinity is a periodic trend increasing (becoming more exothermic) to the right and up.
13. **C is correct.** The answer we are looking for must explain shielding. With each new period, a new shell is added which shields the new electrons from the greater nuclear charge.
14. **A is correct.** If you substitute H for X in the equation for Δ in the passage, you can only arrive at zero.
15. **A is correct.** C and O are close together in electronegativity and will form a covalent bond.

Passage III

16. **B is correct.** Only water is caught in chamber 1. The change in mass of chamber 1, 0.9 grams, is all water. 0.9 grams of water divided by 18 g/mol gives 0.05 mole of water. All the hydrogen came from the sample, and all the oxygen came from the excess oxygen. For every mole of water, there are 2 moles of hydrogens, so there is 0.05 x 2 = 0.1 mole of hydrogen in the sample. Doing the same with the carbon dioxide caught in chamber 2 we have: 4.4/44.2 = 0.1 of CO_2, or 0.1 mole of carbon from the sample. This is a 1:1 ratio. The empirical formula is CH.
17. **B is correct.** The molarity of O_2 is equal to the molarity of the welding gas or any other ideal gas at the same temperature and pressure. Density divided by molecular weight is molarity. Therefore, we can set the ratios of density to molecular weight for oxygen and the welding gas equal to each other. We get: 1.3/32 = 1.1/M.W.

Explanations

18. **D is correct.** CH_2O has a molecular weight of 30 g/mol. Thus, we must multiply this by 4 to get 120. So, for the molecular formula we need four times as many atoms of each element from the empirical formula.
19. **B is correct.** O_2 is the limiting reagent. Only 0.5 mole of propane can react, producing 2 moles of water.
20. **A is correct.** The passage says that $CaCl_2$ absorbs water. Thus if chamber 2 were in front of chamber 1, it would weigh more because it would absorb both water and carbon dioxide. The amount of carbon dioxide is calculated from the weight of chamber 2, so the calculated value would be too high.
21. **D is correct.** All of the welding gas must be reacted because the mass of the original sample is divided by the moles of carbon and hydrogen to find the molecular weight. If all the gas were not reacted, the calculated molecular weight would be too large. Adding excess oxygen ensures that all of the welding gas reacts.

Stand Alones

22. **B is correct.** Chlorine takes on an additional electron to become an ion.
23. **D is correct.** This is the Heisenberg uncertainty principle.

Explanations to Chemistry In-class Exam 2

Passage I

24. **B is correct.** The definition of $Z = PV/(nRT)$ is always 1 for an ideal gas.
25. **A is correct.** If a and b are both zero, the van der Waals equation becomes $PV = nRT$, the ideal gas law.
26. **C is correct.** You *can* figure this out from the passage, but it's a lot easier to fall back on your previous knowledge: gases behave most ideally at high temperature and low pressures.
27. **C is correct.** Volume is *inversely* proportional to pressure. **A:** $K.E. = 3/2\ kT$. **B**: $PV = nRT$, **D:** This is one of the assumptions underlying the derivation of the ideal gas law.
28. **B is correct.** Condensation is due to intermolecular attractions, which are neglected for ideal gases. For **D**, start with $K.E. = 3/2kT$. Then $1/2mv^2 = 3/2kT$, so v is proportional to the square root of T.
29. **C is correct.** Equations involving products or ratios of temperature are meaningless if the zero of the temperature scale is not absolute zero. **A** and **B** are true statements, but they don't explain why absolute temperature must be used.

Passage II

30. **B is correct.** Many organic chemistry reactions are base-catalyzed. Even if you didn't know this, you should have made the assumption that if Acids can act as homogeneous catalysts, then so can bases. Unlike the equilibrium constant, the rate constant can be affected by factors such as catalysts, surface area of the reactants, and even stirring, so **A** is incorrect. **C** is incorrect because catalysts affect rate, not equilibrium.
31. **D is correct.** **B** and **C** are intermediates (produced then consumed) not catalysts (consumed then regenerated). **A** is a reactant.
32. **A is correct.** See the Arrhenius equation:

$$k = zpe^{-Ea/RT}$$

You should also memorize the fact that temperature always increases the rate of a reaction. Even in the case of biologically catalyzed reactions, heat increases the reaction rate until the enzyme is denatured. Once the enzyme is denatured, although the reaction rate slows, the reaction takes a new pathway, and is no longer the same reaction.

33. **A is correct.** If a catalyst only affected the rate in one direction, the equilibrium would be affected. A catalyst doesn't change the equilibrium. This can also be seen from a reaction profile diagram as shown in question 37.
34. **D is correct.** The catalyst is not necessarily the only factor influencing pH.
35. **D is correct.** Choice I is seen from the standard form of the rate law: rate = [A][B]. For choice II, imagine the saturation kinetics exhibited by enzyme catalysts:

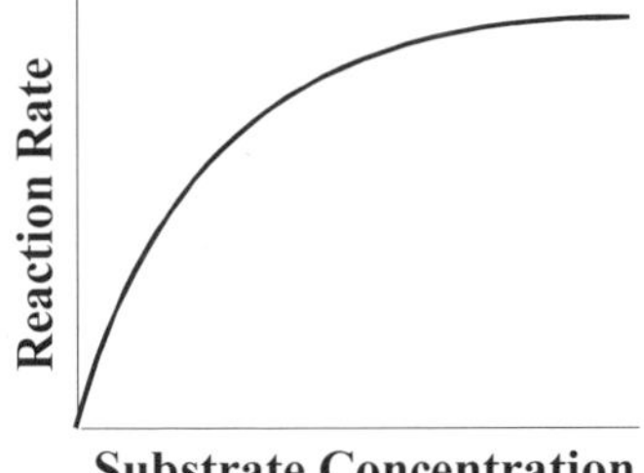

This should make it clear that the ratio of concentrations of the catalyst and the substrates affect the rate of a reaction. This ratio can be changed by changing the concentration of the catalyst. Thus the concentration of a catalyst can affect the rate of a reaction. For choice III, a heterogeneous catalyst is one that is not in the same phase as the reactants. Increasing the surface area of a heterogeneous catalyst is like increasing the concentration. The reaction is affected for the same reasons as in choice II. The reason that a heterogeneous catalyst is typically in the form of metal shavings as opposed to a solid metal bar is to increase surface area. Choice IV you should know from the Arrhenius equation: $k = zpe^{-Ea/RT}$.
36. **C is correct.** The MCAT sometimes uses the phrase "van der Waals" forces as a synonym for London Dispersion Forces. A more modern meaning is as a synonym for intermolecular forces. In either case, this is a correct answer. Hydrogen bonding requires a hydrogen atom bonded to a nitrogen, fluorine, or oxygen. **D** is from an episode of Star Trek.
37. **B is correct.** Only the activation energy is changed by a catalyst. The initial and final states are not affected!

Passage III

In this experiment, Reaction 2 uses up I_3^- as it is formed. When all the $S_2O_3^{2-}$ is used up in Reaction 2, the I_3^- reacts with the starch to turn black. The black color signals the experimenter that all the $S_2O_3^{2-}$ is used up. The experimenter now knows that half as much I_3^- was used up in the same time, and can calculate the rate for Reaction 1. This depends upon Reaction 2 being the fastest reaction.

38. **A is correct.** If we look at Reactions 1 and 2 as two steps of a single reaction, we know that the rate of the slow step is equal to the rate of the overall reaction. Equation 1 measures the time necessary for a specific number of moles of I_3^- to be used by Reaction 2. (Notice that the rate of change of ½$[S_2O_3^{2-}]$ will be equal to the rate of change of $[I_3^-]$) If Reaction 2 were not the fast step, then Equation 1 would not measure the rate of Reaction 1 accurately. Since Reaction 2 is the fast step, the time *t* required to use up ½$[S_2O_3^{2-}]$ is equal to the time needed to produce $[I_3^-]$. The $[I_3^-]$ concentration produced divided by the time necessary to produce it is the rate of Reaction 1. Equation 1 is not derivable from the rate laws of Reactions 1 and 2.
39. **A is correct.** A temperature decrease reduces rate and makes the reaction take longer.
40. **D is correct.** The rate law is found by comparing the rate change from one trial to the next when the concentration of only one reaction is changed. Comparing trials 1 and 2, when the concentration of I^- is reduced by a factor of two, the rate is also reduced by a factor of two. This indicates a first order reaction with respect to I^-. D is the only possible answer.
41. **B is correct.** The exponents in the rate law indicate the order of the reaction with respect to each concentration.
42. **D is correct.** The starch is used to measure the rate of Reaction 1, and does not affect the rate. Although C is true, it does not answer the question as well as D.

Explanations

43. **C is correct.** Equation 1 gives the rate of Reaction 1. $S_2O_3^{2-}$ is not part of Reaction 1 and its concentration does not change the rate. If rate doesn't change, then, according to Equation 1, t must increase with $S_2O_3^{2-}$.
44. **B is correct.** A catalyst increases the rate of a reaction.

Stand Alones

45. **A is correct.** In a reaction at equilibrium, the rate of change in the concentrations of both products and reactants is zero. This does not mean that the concentrations of reactants and products are equal, nor that the rate constants are equal.
46. **C is correct.** Some of both gases will effuse from side 1 to side 2. This means that the partial pressures of both gases will decrease. (Remember, partial pressure is the pressure of the gas as if it were alone in the container. Thus if we reduce the number of moles of a gas at constant volume and temperature, we reduce its partial pressure.) Since hydrogen will diffuse more rapidly than oxygen, the mole fraction of oxygen will increase.

Explanations to Chemistry In-class Exam 3

Passage I

47. **A is correct.** Nickel is an element and is a solid is in its natural state at 298 K. Thus, the enthalpy of formation of solid nickel at 298 K is zero.
48. **C is correct.** Chemist A chooses the direction of the reaction based upon chemical stability, and then says that the direction will change at higher temperatures. This is tantamount to saying that the stability will switch at higher temperatures, which, by the way, is also correct. D is, of course, a false statement. The entropy shown for Reaction 1 is the entropy of the reaction. In other words, it is the entropy of the system, and not the entropy of the universe. Chemist A's statement is correct, and does not contradict the second law of thermodynamics. A and B are contradicted because Chemist A says the direction is temperature dependent.
49. **B is correct.** Use Le Chatelier's principle. Read the first sentence of the passage carefully, and notice that to purify nickel, the reaction must move to the left. There are four gas molecules on the left side of the reaction and only one on the right. Pressure pushes the reaction to the right. The reaction is exothermic when moving to the right, so high temperature pushes the reaction to the left.
50. **C is correct.** The reaction is the system, and everything outside the reaction makes up the surroundings. The entropy change given in the passage refers to the system not the universe. The second law of thermodynamics says that a reaction is spontaneous when the entropy of the universe is positive. The entropy of the system may be positive or negative. A and B are false statements. D is a false statement as well. The reaction runs until the entropy of the universe is maximized.
51. **A is correct.** Use $\Delta G = \Delta H - T\Delta S$. Don't forget to convert J/K to kJ/K.
52. **D is correct.** Spontaneity is dictated by Gibbs energy. 1 atmos. is standard state for a gas, so $\Delta G = \Delta G^o$. When Gibbs energy is negative, a reaction is spontaneous. If enthalpy change is negative and entropy change is positive, then Gibbs energy change must be negative. You can use $\Delta G^o = \Delta H^o - T\Delta S^o$. Check this as follows: If the partial pressures are 1, then the reaction quotient Q is 1, and the log of the reaction quotient is zero. From the equation $\Delta G = \Delta G^o + RT\ln Q$ we see that $\Delta G = \Delta G^o$. The reaction is spontaneous.

Passage II

Note: A heat engine obeys the first law of thermodynamics. It must expel the same amount of energy as it takes in.

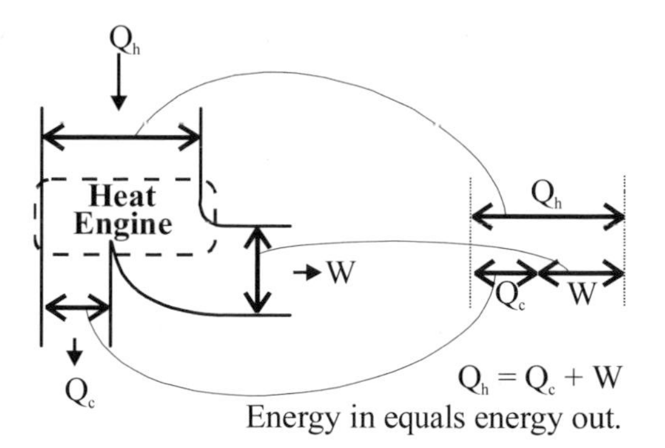

53. **C is correct.** If $Q_h' = Q_h$ and $Q_c' = Q_c$ then $W' = W$. Thus the efficiencies of the engines must be equal. Since only a Carnot engine can be as efficient as another Carnot engine, Engine 1 must be a Carnot engine.
54. **C is correct.** If the all the work done by Engine 1 is done on Engine 2, the net work is zero. Since Engine 1 is not a Carnot engine, the entire process is not reversible. The result of any nonreversible process where no work is done must be that heat energy is transferred from the hot reservoir to the cold reservoir. Engine 1 has a lower e than Engine 2 and thus requires more heat energy to create as much work. According to conservation of energy, this extra heat energy input must be matched by extra heat energy output.
55. **D is correct.** Engine 2 is a Carnot engine, and, as the passage states, it has the highest possible efficiency of any engine working between the existing heat reservoirs. Thus only a change in the heat reservoirs will increase its efficiency. **The answer must therefore be C or D.** For greatest efficiency we want to remove the most heat energy possible from the hot reservoir and expel the least amount possible to the cold reservoir thus getting the most work with the least amount of wasted energy. Removing heat energy from the hot reservoir decreases its entropy, while adding heat energy to the cold reservoir increases its entropy. As the temperature of the hot reservoir increases, removing heat energy has less effect on the change in entropy, so more heat energy can be removed. The reverse is true for the cold reservoir. Since, in a Carnot engine, the change in entropy must be zero, the extra heat energy removed from the hot reservoir must be added to work. The engine becomes more efficient. Thus maximizing the temperature difference increases efficiency. This can be derived from the equation in the passage (the long method) as follows. Considering magnitudes only we have:

$$Q_h/T_h = Q_c/T_c$$
$$Q_c/Q_h = T_c/T_h$$

$$e = W/Q_h$$
$$e = (Q_h - Q_c)/Q_h$$
$$e = 1 - Q_c/Q_h$$

substituting T_c/T_h for Q_c/Q_h we have:

$$e = 1 - T_c/T_h.$$

56. **C is correct.** In a non-cyclical quasi-static process (a process always infinitely close to equilibrium), it is possible to convert heat energy to work with 100 percent efficiency.
57. **C is correct.** The exhaust is wasted energy. $Q_h = W + Q_c$. $P = Q_h/t = W/t + Q_c/t = (mgh)/\ 20$ s + 1000 J/s = 4000/20 + 1000 = 1200 J.
58. **A is correct.** $PV = nRT = (m/\text{M.W.})RT$. $P = 1000 \times 8.314 \times 400/4 = 8.314 \times 10^5$ Pa. By the way, if the gas did not behave ideally, the real pressure would be lower. There is no answer lower than A, and the gas does behave very nearly ideally because it is at high temperature.

Passage III

59. B is correct. The transition state corresponds to the top of the energy curve.

60. C is correct. The energy of activation is given by the vertical displacement from the reactants to the top of the energy curve.

61. D is correct. The smallest energy of activation is the most kinetically favored.

62. A is correct. The largest drop in energy is the most thermodynamically favored.

63. A is correct. A catalyst lowers the energy of activation but does not change the energy difference between the reactants and products.

64. B is correct. The change in energy is energy of products minus reactants.

65. A is correct. The kinetically favored product is the one with a lower energy of activation. The difference in their equilibrium is due to conflicting thermodynamics and kinetics. At a low temperature (T_2), the thermodynamically favored product does not have enough energy to reach the activated complex, so no reaction occurs. The kinetically favored reaction does reach the activated state and a reaction can occur. At the high temperature (T_1) both reactions occur but the reverse of the thermodynamically favored occurs only with a relatively lower probability. Thus the thermodynamically favored reaction predominates. This is not always true but is a concept of which you must be aware.

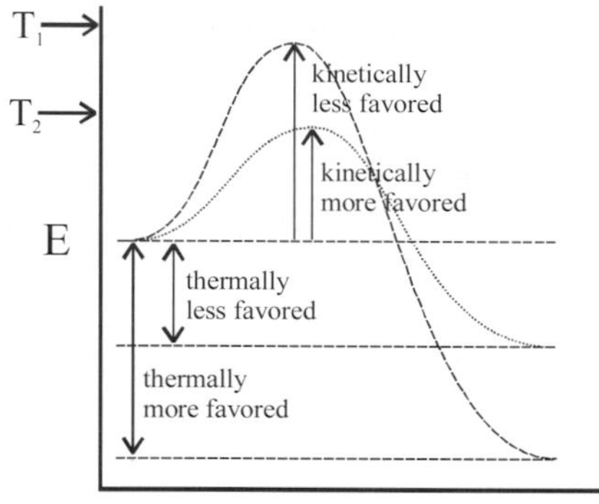

Stand Alones

66. B is correct. Transfer by contact is conduction.

67. C is correct. The second law of thermodynamics says that entropy of the universe increases for any process. By straightening up his room, the man increased the order in his room, and thus decreased its entropy. In order for the entropy of the universe to have increased, there must be a larger increase in entropy of the surroundings. Only C provides an explanation for this.

68. C is correct. The entropy of the system is equal to change in entropy of the two reservoirs. $\Delta S = Q/T$ for each reservoir. The change in entropy of the first reservoir is negative because heat energy is leaving the system ($-1000/200 = -5$), and the change in entropy of the second reservoir is positive because heat energy is entering the system ($1000/100 = 10$). The sum of the two entropy changes is +5. You should have at least narrowed down the possibilities to C and D because the change in entropy for any isolated system must be positive for any process.

69. D is correct. To answer this question, you should realize that the temperature of an ideal gas is a function of its internal energy only. An ideal gas has no forces between molecules, so there is no potential energy between molecules. Thus, all molecular energy can be considered kinetic ($K.E. = 3/2kT$). The question says that gas A doesn't change temperature. Since gas B does not exchange energy with its surroundings, the kinetic energy of the molecules is conserved, and the gas does not change temperature. Thus, for both gases, only P changes when V changes ($PV = nRT$). The question states that the change in volume is identical for both gases, so the pressure is identical for both gases. Enthalpy is $U + PV$. The change in U is zero for both gases since temperature is constant, and the change in P and V is the same for both gases. Thus enthalpy change is equal for both gases. [A simple way of stating all of this is: For an ideal gas both internal energy and enthalpy are functions of temperature only.] **Warning:** This is true only for an ideal gas.

Explanations to Chemistry In-class Exam 4

Passage I

70. C is correct. The first experiment shows that no nitrite was in the solution. Had there been nitrate, nitrogen bubbles would have formed as per the reaction, and then barium sulfate would have precipitated upon the addition of barium chloride. The second experiment demonstrates that nitrates exist. The water in the moistened litmus paper reacted with ammonia gas to make OH^- ions turning the paper blue. Ammonia gas resulted from the reaction of nitrate with Devarda's alloy.

71. B is correct. Aluminum begins in its elemental form, so it has an oxidation state of zero. It then loses three electrons, so it is oxidized to +3. Each OH^- has an oxidation state of –1. The $Al(OH)_4^-$ molecule has a 1– charge so Al must neutralize three of the four negatives from the OH^- ions. Oxidation state is covered in Lecture 7.

72. B is correct. The precipitate would result when the sulfate ion from the reaction reacts with the barium ion to form barium sulfate. The bubbles would be created by the formation of nitrogen gas.

73. B is correct. Since NaOH totally dissociates, the OH concentration is 6 *M* (between 1 *M* and 10 *M*), thus the pOH is between 0 and –1; thus the pH must be between 14 and 15. This corresponds to an H^+ ion concentration greater than $1\text{x}10^{-15}$ *M* but less than $1\text{x}10^{-14}$ *M*. Acids and bases are covered in Lecture 6.

74. C is correct. Change in temperature can change the rate constant and the equilibrium constant but it can only increase the rate of the reaction.

75. D is correct. The litmus paper is turned blue when the basic ammonia gas from the reaction in the experiment reacts with the water in the paper.

Passage II

76. A is correct. This is definitional: products over reactants excluding pure liquids and solids.

77. C is correct. K_{sp} is a constant; solubility of $Ca(IO_3)_2$ is not. By Reaction 2, as acidity increases, IO_3^- ions are used up, pulling Reaction 1 to the right. Or just look at Table 1. If you want to see why you can't just use Le Chatelier's principle on Reaction 2, simply add the two equations together to get:

$$Ca(IO_3)_2 + H^+ \rightarrow Ca^{2+} + IO_3^- + HIO_3$$

Now when you add H^+ to this equation, it moves to the right, dissolving $Ca(IO_3)_2$.

78. B is correct. You should recognize the $y = mx + b$ form of the equation. This is the equation of a line. The *b* in this case is not zero, but is $(K_{sp})^{½}$.

79. B is correct. You must realize that the new solution is no longer saturated. These means that Reaction 1 is not in equilibrium. No precipitate exists. New Ca^{2+} ions do not immediately create precipitate because the solution is not saturated. There is no leftward shift because there is no equilibrium. Thus A, C and D are wrong. Iodic acid is in equilibrium however. Increasing H^+ ions shifts Reaction 2 to the right, creating more HIO_3. This is, of course, why Solution 1 is no longer saturated after adding the acid.

80. C is correct. The solid is filtered off, so more acid cannot dissolve more solid.

81. D is correct. The net equation is $Ca(IO_3)_2(s) + H^+(aq) \rightarrow Ca^{2+}(aq) + IO^{3-}(aq) + HIO_3(aq)$. CaOH in aqueous solution will produce both OH^- and Ca^{2+}. The OH^- will reduce the H^+ in solution. Since the solution is saturated (in equilibrium), Le Chatelier's principle predicts that both of these changes will shift the reaction to the left producing $Ca(IO_3)_2(s)$.

82. C is correct. Reaction 2 will shift left via Le Chatelier's principle. The resulting increase in IO_3^- will shift Reaction 1 to the left due to the common ion effect, creating precipitate in the already saturated solution. The H^+ ion concentration in the formula for solubility is from hydrogen ions in solution before iodic acid is added. If a different acid were added (like HCl), the H^+ ion concentration would move Reaction 2 to the right and thus Reaction 1 to the right.

Passage III

A ligand is an ion or neutral molecule that can donate a pair of electrons to form a coordinate covalent bond with a metal ion. EDTA is a chelating agent (a ligand that makes more than one bond to a single metal ion). It wraps around its metal ion like a claw. *Chele* (χηλη) means claw in Greek.

83. **D is correct.** The passage states that EDTA reacts with other metal ions. If more EDTA is used up, the scientist will assume that it is being used up by calcium and magnesium ions. This will result in an overestimation of these ions.
84. **C is correct.** EDTA is donating a pair of electrons in a coordinate covalent bond, so it is a Lewis Base.
85. **A is correct.** This is the common ion effect on the K_{sp}. The chlorine ion will remove some of the Ca^{2+} pulling the reaction to the right.
86. **C is correct.** The passage states that there is a one-to-one stoichiometry between EDTA and its metal ion. (50 mL)(0.001 mol/L) = (25 mL)x. x = 0.002 mol/L
87. **D is correct.** D is the best explanation. Under high pH conditions, protons are stripped from the carboxylic acids allowing the ligand to bond to the cation. Indicators change color over a range so C is wrong. Calcium carbonate dissolves in an acid solution so B is wrong. A is irrelevant.
88. **A is correct.** You should know that Na^+ is very soluble, and H^+ does not form mineral deposits.
89. **A is correct.** The association constant from Table 1 is 5×10^8. The association reaction is:

$$Mg^{2+} + EDTA \rightarrow EDTA\text{-}Mg^{2-}.$$

The $K_{assoc} = 5 \times 10^8 = [EDTA\text{-}Mg^{2-}]/[EDTA][Mg^{2+}]$

Since half the magnesium is bound, $[EDTA\text{-}Mg^{2-}]$ is 9 ppm, which is 5×10^{-4}. The remaining half of $[Mg^{2+}]$ is 9 ppm, which is also 5×10^{-4}. Plugging these into the equilibrium expression leaves the remaining concentration [EDTA] = $1/(5 \times 10^8)$ or 2×10^{-9}.

Stand Alones

90. **C is correct.** Water boils at 100 °C and atmospheric pressure, 760 torr. Boiling point is where vapor pressure meets atmospheric pressure. Thus water vapor pressure must be below 760 at 95 °C. But it must also rise with increasing temperature, so it must be above 23.8 torr.
91. **B is correct.** No calculations are required since the vapor pressure would be somewhere between the vapor pressures of the pure liquids. The solution follows Raoult's law.
92. **B is correct.** Since the reaction was exothermic, the vapor pressure deviated negatively from Raoult's law. Depending upon the ratios of the liquids in solution, the vapor pressure could be lower than either or just lower than B. (A has a higher boiling point thus a lower vapor pressure.) The boiling point must have gone up from B because the vapor pressure went down from B.

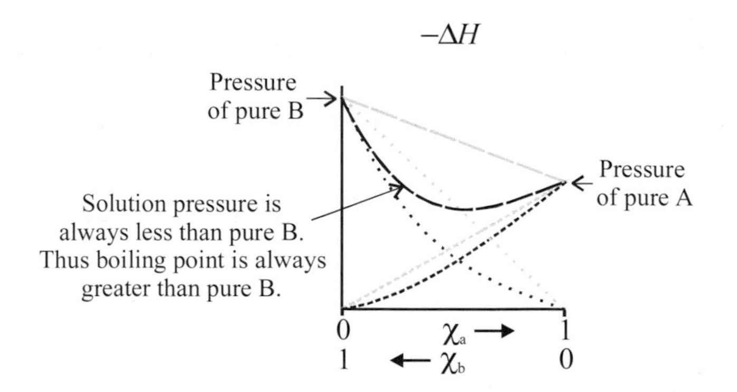

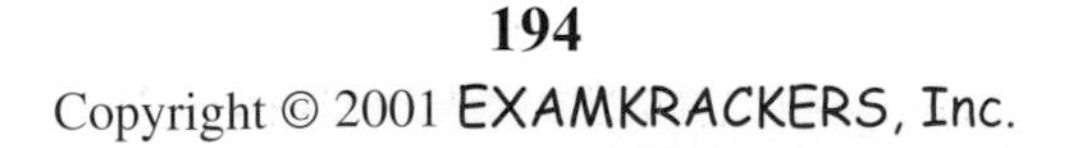

Explanations to Chemistry Practice Exam 5

Passage I

93. C is correct. The salt lowers the freezing point of water. This is necessary in order to insure that the water can bring the unknown solution to its freezing point.

94. C is correct. Even if the others completely dissociate, NaCl still releases more particles per gram than the others. (2 particles/58 grams) (A = 3/172, B = 2/120, and D = 3/115)

95. C is correct. The copper stirrer acts to evenly distribute the heat throughout the solution by convection.

96. B is correct. Since the freezing point depression was lower for Student 1, there must have been fewer particles for the same amount of mass. Thus Student 1 had an unknown with lower molecular weight.

97. D is correct. The freezing point of pure water can only go down when a solute is added, and Student 2's solution has a higher freezing point than zero. From the K_f of cyclohexane we know that since student 1's solution did not drop by 20.2 °C, the normality of his solution is less than one. Thus an aqueous solution will have a freezing point greater than –1.86. (The K_f of water is 1.86.)

98. D is correct. We use $\Delta T = K_f m$:

$$m = (\text{grams}_{\text{solute}}/\text{M.W.})/(\text{volume}_{\text{solvent}} \times \text{density}_{\text{solvent}} \times \text{kg/gram})$$

plugging into $\Delta T = K_f m$ and rearranging we have:

$$\text{M.W.} = (K_f)\,(\text{grams}_{\text{solute}}) / \{(\Delta T)\,(\text{volume}_{\text{solvent}} \times \text{density}_{\text{solvent}} \times \text{kg/gram})\}$$

$K_f = 20.2$, $(\text{grams}_{\text{solute}}) = 0.5$, $\Delta T = (6.6 - -3.5) = 10.1$, $\text{volume}_{\text{solvent}} = 10.0$, $\text{density}_{\text{solvent}} = 0.78$, kg/gram = 1/1000

Thus

$$\text{M.W.} = (K_f)\,(\text{grams}_{\text{solute}}) / \{(\Delta T)\,(\text{volume}_{\text{solvent}} \times \text{density}_{\text{solvent}} \times \text{kg/gram})\}$$

99. D is correct. D is the only nonvolatile solute that is soluble in cyclohexane.

100. B is correct. The only explanation is B. As long as some solution remains liquid, the energy removed by the ice bath creates bonds forming a solid. As soon as the entire solution is frozen, the energy removed from solution lowers the temperature.

Passage II

101. A is correct. The acid and base are totally dissociated to begin with. This reaction takes high energy molecules and makes a low energy molecule, releasing heat.

102. B is correct. Heat is released.

103. A is correct. 40 mL x 1 cal/°C mL x –3.3 °C = –132 cal. Since heat is released, we already know the answer is negative.

104. B is correct. The ammonium nitrate would require energy to dissociate before releasing energy to form water.

105. D is correct. Heat per mole is an intensive property.

106. D is correct. The temperature of the calorimeter is higher than the surroundings throughout the experiment. Heat always moves from hot to cold.

Passage III

107. A is correct. The negative slope on the phase diagram demonstrates that water expands when freezing.

108. C is correct. The line between A and E represents equilibrium of gas and solid.

Explanations

109. D is correct. Point D is the triple point.

110. D is correct. Point C is the critical point.

111. B is correct. The negative slope between the solid and liquid phases of water in Figure 2 represents melting point at different temperatures and pressures. As pressure increases, the temperature decreases moving along the line.

112. A is correct. The normal boiling point is the boiling point at local atmospheric conditions (1 atm). The triple point must be at or below the temperature and pressure of the normal boiling point. Only A fulfills these requirements.

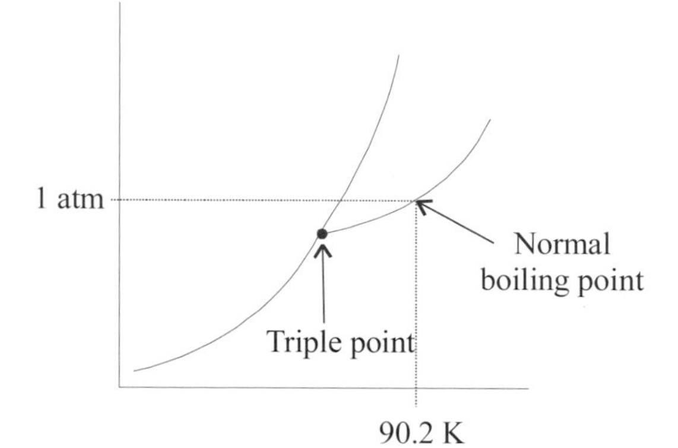

113. C is correct. See the graph.

Stand Alones

114. A is correct. During a phase change, temperature, and thus molecular kinetic energy, is constant. Breaking bonds always absorbs energy. Ice cools things when it melts.

115. A is correct. The potential energy of the water at the top of the falls becomes kinetic energy as it drops, and then thermal energy at the bottom of the falls. Thus $mgh = Q = mc\Delta T$, or $\Delta T = gh/c$.

Explanations to Chemistry In-class Exam 6

Passage I

116. A is correct. This is the common ion effect in Reaction 1.

117. A is correct. For all trials, at the equivalence point there are Ca^{2+} ions, Na^+ and Cl^- ions. (Trial 1 has no Na^+ ions, but these don't affect pH anyway.) The Ca^{2+} started as a saturated solution. As OH^- ions are removed from solution by the acid, there is no precipitation of $Ca(OH)_2$. When all the OH^- ions are neutralized by the acid, the pH is 7 for all trials. The equivalence point is 7 for all trials. For trials with NaOH, the pH begins higher. More HCl is required to neutralize this extra base. The curves are exactly the same, but the trial one curve starts at a lower pH and requires less HCl.

118. B is correct. This is just the normal K_{sp}.

119. D is correct. $pOH = -\log[OH^-]$ In this case, the significant OH^- ion contribution comes completely from NaOH which dissociates completely. $1 > pOH > 2$ The pH is $14 - pOH$ = between 12 and 13.

120. D is correct. The $Ca(OH)_2$ solution begins as basic, and when the $Ca(OH)_2$ is totally dissolved, it should be neutral. See answer to question #117.

121. A is correct. This is the titration of a strong base with a strong acid. See question #117.

Passage II

122. C is correct. Since the K_w is higher, the hydrogen ion concentration will be higher also. But it would have to be 10 times higher in order for the pH to be higher.

123. C is correct. Set K_w equal to 1 mole of hydrogen ions times the OH concentration.

124. B is correct. As T increases, the hydrogen ion concentration increases.

125. C is correct. A proton is lost to form the conjugate base of a Bronsted acid.

126. C is correct. The leveling effect in water occurs because water readily accepts all protons from both acids. The equilibrium in water is so far to the right for both reactions that no comparison can be made. Although acetic acid accepts protons from both HCl and $HClO_4$, it does not do so as readily as water (it is a weaker proton acceptor or Bronsted-Lowry base). Thus, an equilibrium is established for both reactions, and the equilibriums can be compared.

127. A is correct. Definitional.

Passage III

128. C is correct. The pH must be where both indicators can have the proper color.

129. C is correct. The lower pK_a of sulfuric acid demonstrates that it is a stronger acid than sulfurous acid. The second proton is not the major contributor to the acid strength, so D is wrong.

130. A is correct. The limestone would buffer the lake. A lake with a large surface area would collect more acid rain; a deep lake would have more water with which to dilute the acid.

131. C is correct. The conjugate base is found by removing a proton.

132. A is correct. Minus eight for the oxygens, plus two for the hydrogens leaves minus six which must be counter balanced. Minus six for the oxygens, plus two for the hydrogens leaves minus four which must be counter balanced.

133. C is correct. We could never raise the pH by adding an acid. Water is the main contributor of H^+. To find the pH, we add the 5×10^{-8} ions contributed by H_2SO_4 to the 1×10^{-7} ions contributed by water. This leaves 1.5×10^{-7} H^+.

134. D is correct. This is the titration of a diprotic acid with a strong base.

Stand Alones

135. A is correct. Perchloric acid is the strongest acid, thus it has the weakest conjugate base. In oxy acids, the more oxygens, the greater the acid strength.

136. C is correct. This is the definition of the half equivalence point

137. A is correct. Use the Henderson-Hasselbalch equation. $pH = pK_w - \log(A^-/HA)$ => $pH = \log(1.8 \times 10^{-7}) - \log(1/10)$ => $pH = 4.7 - 1 = 3.7$

138. B is correct. The conjugate acid has the K_a that equals K_w/K_b.

Explanations to Chemistry In-class Exam 7

Passage I

139. C is correct. $Cr_2O_7^{2-}$ is reduced to Cr^{3+}. Although this doesn't look like reduction from the charges, Cr in the dichromate has an original oxidation state of +6.

140. C is correct. This is simply reading the graph.

141. B is correct. For each mole of dichromate that is reduced, 6 moles of Fe are oxidized. The top portion of B gives the number of moles of dichromate reduced times six, which is the number of moles of Fe oxidized. (The equivalence point is where all the iron has been oxidized.) We divide this by the original volume of Fe solution and get the molarity.

142. D is correct. This is reading comprehension from the last paragraph.

143. A is correct. This is the definition of the equivalence point for an oxidation-reduction titration.

144. C is correct. Oxygen is –2. There are 7 oxygens which make –14. The 2– charge on the ion takes care of 2 of the 14 negatives. The 2 chromiums must take care of the other 12. That's +6 for each chromium.

Passage II

145. A is correct. The half reactions must be rearranged in such a fashion so as the total voltage is positive (meaning the battery is discharging). This requires reversing the top half reaction. When reversed, this reaction becomes oxidation, which takes place at the anode.

146. D is correct. When we recharge the battery, the reactions are both reversed from the positions in question one.

147. C is correct. Ni is being reduced, so NiO_2 is the oxidizing agent. The compound with many oxygens is often the oxidizing agent.

148. A is correct. The cell has a negative potential and is forced to run in the nonspontaneous direction.

149. C is correct. The two half reaction potentials must be added after they have been rearranged to represent the galvanic cell. This means that the first half reaction is reversed. If this potential is applied, the cell can be recharged back to this potential which is the standard potential.

150. B is correct. The passage tells us that sodium is at the anode so it is oxidized. Sodium is not normally oxidized to a +2.

151. D is correct. Current moves opposite to electrons. Since electrons flow from Half Reaction 1 to 2, current flows from Half Reaction 2 to 1.

Passage III

152. D is correct. The reaction quotient is in the same form as the equilibrium constant. Pure solids and liquids should not be used in the law of mass action to solve for the equilibrium constant.

153. A is correct. Plug the numbers into the Nernst equation. $(10^2)^2$ is 10^4. The log of 10^4 is 4. Thus the potential is negative and the voltage is twice 0.0592 V.

154. B is correct. See the last equation in the passage.

155. B is correct. See the last equation in the passage. A 1 *M* solution of HCl has a pH of 0.

156. A is correct. An increase of H^+ by a factor of 10 is a decrease in pH of 1. $\mathcal{E} = 0.059\ (pH_{test} - pH_{reference})$ Each decrease in $pH_{reference}$ of 1 amounts to an increase in $\mathcal{E}$ of 0.059 V or 59 mV.

157. A is correct. The current will try to even the charges in the solutions. Since we have more positive charge on the concentrated side, the current moves to the less concentrated side.

158. B is correct. Again, current flows toward the less positive side, which is the basic side, which has less H^+ ions.

Stand Alones

159. D is correct. You should know this one reduction potential which is: $2H^+ + 2e^- \rightarrow H_2$ $\mathcal{E} = 0$. When this is added to the oxidation of solid zinc, the potential is positive, which means spontaneous.

160. C is correct. At equilibrium, there can be no potential; neither direction of the reaction is favored.

161. B is correct. The products and reactants are at standard state, and therefore their potential defines the standard potential $\mathcal{E}^o$. A is wrong because they are not at equilibrium when $Q = 1$. C is wrong because the potential is positive. D is wrong because Q is at 1 and Q will move toward K. The reaction is spontaneous from here so products will increase, and Q will increase. Therefore, K must be greater than 1.

About the Author

Jonathan Orsay is uniquely qualified to write an MCAT preparation book. He graduated on the Dean's list with a B.A. in History from Columbia University. While considering medical school, he sat for the real MCAT three times from 1989 to 1996. He scored in the 90 percentiles on all sections before becoming an MCAT instructor. He has lectured in MCAT test preparation for thousands of hours and across the country for every MCAT administration since August 1994. He has taught premeds from such prestigious Universities as Harvard and Columbia. He was the editor of one of the best selling MCAT prep books in 1996 and again in 1997. Orsay is currently the Director of MCAT for Examkrackers. He has written and published the following books and audio products in MCAT preparation: "Examkrackers MCAT Physics"; "Examkrackers MCAT Chemistry"; "Examkrackers MCAT Organic Chemistry"; "Examkrackers MCAT Biology"; "Examkrackers MCAT Verbal Reasoning & Math"; "Examkrackers 1001 questions in MCAT Physics", "Examkrackers MCAT Audio Osmosis with Jordan and Jon".

An Unedited Student Review of This Book

The following review of this book was written by Teri R---. from New York. Teri scored a 43 out of 45 possible points on the MCAT. She is currently attending UCSF medical school, one of the most selective medical schools in the country.

The Examkrackers MCAT books are the best MCAT prep materials I've seen-and I looked at many before deciding. The worst part about studying for the MCAT is figuring out what you need to cover and getting the material organized. These books do all that for you so that you can spend your time learning. The books are well and carefully written, with great diagrams and really useful mnemonic tricks, so you don't waste time trying to figure out what the book is saying. They are concise enough that you can get through all of the subjects without cramming unnecessary details, and they really give you a strategy for the exam. The study questions in each section cover all the important concepts, and let you check your learning after each section. Alternating between reading and answering questions in MCAT format really helps make the material stick, and means there are no surprises on the day of the exam-the exam format seems really familiar and this helps enormously with the anxiety. Basically, these books make it clear what you need to do to be completely prepared for the MCAT and deliver it to you in a straightforward and easy-to-follow form. The mass of material you could study is overwhelming, so I decided to trust these books--I used nothing but the Examkrackers books in all subjects and got a 13-15 on Verbal, a 14 on Physical Sciences, and a 14 on Biological Sciences. Thanks to Jonathan Orsay and Examkrackers, I was admitted to all of my top-choice schools (Columbia, Cornell, Stanford, and UCSF). I will always be grateful. I could not recommend the Examkrackers books more strongly. Please contact me if you have any questions.

Sincerely,
Teri R----

About the Author

Jonathan Orsay is uniquely qualified to write an MCAT preparation book. He graduated on the Dean's list with a B.A. in History from Columbia University. While considering medical school, he sat for the real MCAT three times from 1989 to 1996. He scored in the 90 percentiles on all sections before becoming an MCAT instructor. He has lectured in MCAT test preparation for thousands of hours and across the country for every MCAT administration since August 1994. He has taught premeds from such prestigious Universities as Harvard and Columbia. He was the editor of one of the best selling MCAT prep books in 1996 and again in 1997. Orsay is currently the Director of MCAT for Examkrackers. He has written and published the following books and audio products in MCAT preparation: "Examkrackers MCAT Physics"; "Examkrackers MCAT Chemistry"; "Examkrackers MCAT Organic Chemistry"; "Examkrackers MCAT Biology"; "Examkrackers MCAT Verbal Reasoning & Math"; "Examkrackers 1001 questions in MCAT Physics", "Examkrackers MCAT Audio Osmosis with Jordan and Jon".

An Unedited Student Review of This Book

The following review of this book was written by Teri R---. from New York. Teri scored a 43 out of 45 possible points on the MCAT. She is currently attending UCSF medical school, one of the most selective medical schools in the country.

The Examkrackers MCAT books are the best MCAT prep materials I've seen-and I looked at many before deciding. The worst part about studying for the MCAT is figuring out what you need to cover and getting the material organized. These books do all that for you so that you can spend your time learning. The books are well and carefully written, with great diagrams and really useful mnemonic tricks, so you don't waste time trying to figure out what the book is saying. They are concise enough that you can get through all of the subjects without cramming unnecessary details, and they really give you a strategy for the exam. The study questions in each section cover all the important concepts, and let you check your learning after each section. Alternating between reading and answering questions in MCAT format really helps make the material stick, and means there are no surprises on the day of the exam-the exam format seems really familiar and this helps enormously with the anxiety. Basically, these books make it clear what you need to do to be completely prepared for the MCAT and deliver it to you in a straightforward and easy-to-follow form. The mass of material you could study is overwhelming, so I decided to trust these books--I used nothing but the Examkrackers books in all subjects and got a 13-15 on Verbal, a 14 on Physical Sciences, and a 14 on Biological Sciences. Thanks to Jonathan Orsay and Examkrackers, I was admitted to all of my top-choice schools (Columbia, Cornell, Stanford, and UCSF). I will always be grateful. I could not recommend the Examkrackers books more strongly. Please contact me if you have any questions.

Sincerely,
Teri R----